Tornado

Wright Aero's Last Liquid-Cooled Piston Engine

Kimble D. McCutcheon

Weak Force Press Huntsville, AL

Weak Force Press
P.O. Box 11815
Huntsville, AL 35814

Printed in the United States of America

01 02

ISBN 0-9710847-0-X

Library of Congress Control Number: 2001097807

Cover Photograph of the Tornado 42, Military Designation R-2160-3 (Arthur Nutt Papers, Acc. 1987-0015, National Air and Space Museum, Smithsonian Institution, SI Neg. No. 2001-2659)

To my mother, Elizabeth, who stands by me as I explore the world.

Contents

Foreword

Having trodden the path taken by Kim McCutcheon, that of writing about aircraft engines, I can appreciate just how difficult a task it is. What makes Kim's accomplishment even more impressive is the fact that he chose to write about one of the most complex aircraft piston engines ever developed. Only a small number of Tornados were built; therefore, unlike engines that were manufactured in the thousands, little documentation survived. But what little information did survive, Kim latched onto to produce this important work.

Even though the Tornado did not get into series production—in fact, it never flew—it is important that this advanced piece of mechanical engineering is saved for future generations to marvel at. Today's engineers will be amazed at the fact that this complex example of the engineer's art was designed with what would be considered antiquated and primitive design tools by today's standards. Instead of computers and computer-aided design tool, these artisans of a bygone age relied on slide rules, intuition, and their own innate feeling of how an engine should be designed. And then whatever broke on the dynamometer was carefully analyzed, as only an experienced engineer schooled in these methods can appreciate, then a design fix was incorporated.

The Tornado exemplified the dogged determination required to bring a complex mechanical design to fruition. Even though one would think that all potential weak points would be accommodated for during the design phase, like any high performance engine, when the rubber hits the road in the form of a prototype engine running on a dynamometer, that's when the real work starts. Kim has done an excellent job of covering these events. Real people, working under arduous conditions, sweated blood and tears getting the Tornado into running condition and then resolving problems as they arose. People are an often overlooked facet of engineering development; Kim McCutcheon has brought these folks into the limelight they so richly deserve. By interviewing the surviving engineers and personnel involved in the Tornado project, a more human touch has been injected into the story. Even so, it's too bad that many of the people mentioned in this book are no longer with us. Engineers are often criticized for poor writing skills; however, it has been my experience that this is not the case. It is thanks to their skilled documentation that it has been possible to relate this fascinating story.

Considering the fact Wright Aeronautical Corporation had a full plate getting desperately needed B-29 engines into some semblance of reliability, along with squeezing more power out of their existing engines, it is remarkable that such an ambitious project as the Tornado was part of those efforts. And then they had their other products to deal with such as propellers and aircraft. These took an inordinate amount of manpower to keep reliability, performance and efficiency where it needed to be and competitive with the competition and Axis powers. Wright was in a fierce uncompromising battle for the aircraft engine business during World War II with Allison and Pratt & Whitney, both of whom were worthy competitors. A similar situation existed with aircraft. Curtiss was constantly updating and improving the P-40 and C-46. Along with these production aircraft, Curtiss developed experimental aircraft in response to Army Air Force bids. The Ascender is a prime example. As if the forgoing was not enough, Curtiss also built aircraft under license, such as the complex Republic P-47.

After many experimental projects were terminated, it would appear at times as if the parties involved wanted to get rid of all the evidence as quickly as possible after cancellation. Thank goodness that, in the case of the Tornado, at least one example survived along with enough documentation to get the ball rolling on this book. It is heart warming to see the entire story related in this wonderful book; from initial design studies to a complete engine, lovingly restored, and residing in a museum.

Graham White

November 1, 2001

Preface

I sit in my rental car looking at the locked gate of a small New Jersey aviation museum. It is a gray November morning, and I am the first to arrive, early by ten minutes. I have traveled a great distance, and am here for one thing—to see the sole surviving example of an obscure experimental aircraft engine. I cannot say precisely what it is about this engine that has captured my imagination. Perhaps it is its complexity. Perhaps it is the fact it has survived against almost insurmountable odds. Perhaps it is that I sense a good story.

One thing about this engine that fascinates me is that it represents so many of the Government projects I have personally witnessed. It was conceived to fill a real need, created by an optimistic and energetic group of engineers, and then abandoned when the political tides turned. Its creators were left with a sense of loss and the feeling that their most precious years had been wasted. The evidence of such projects is usually quickly destroyed, and the memories fade as the cycle repeats with the next new thing. That is why the mere existence of this engine is so exceptional.

The museum volunteers start to arrive. The gate is unlocked, and I park. My tools—cameras, laptop, scanner, and notepads—stay in the car. I don't want to overwhelm the museum volunteers with too many requests first thing in the morning. Besides, I always like to get the feel of a new museum before actually going to work, and the tools will be a distraction.

I greet the volunteer, pay for my admission, and walk into the exhibit area. It is dark, and I am alone. I find the light switch and start to form impressions as the high-pressure sodium lamps warm up and change color. For a small museum, this one is extraordinary. In addition to being clean and neat, it has a disproportionately large collection of rare and exotic engines. There are rocket engines from the Bell X-2 and North American X-15, a beautifully restored Wright R-1820, a two-cylinder Lawrance, and, of course, the engine of my interest, a Wright Tornado.

This is no ordinary place. Evidence of accuracy and attention to detail is everywhere. Unlike the engine collections of many aviation museums, the placards are correct, the engines are authentically restored, and they are prominently displayed. In many museums, one gets the impression that the aviation gods made them take some engines before they could have any airplanes.

After a brief museum walk around, I collect my tools and go to work, taking photographs and scanning documents. I have come here because I am interested in thirteen high powered liquid-cooled engines ordered by the U.S. armed forces between 1932 and 1945. Only one of these engines could be considered a success. The Wright Tornado that is here is probably the most obscure. It was by no means the success.

I am hoping to get enough research material for, at best, a long paper. Little do I know that I am at the beginning of a remarkable story about remarkable people trying to do nearly impossible things. This is the story of an engine that never flew, that was repeatedly saved from the scrap heap by sheer luck, and was finally restored to beautiful condition by a dedicated group of volunteers. Over the next two years, I will get to know many of these people well, and indeed be able to claim several as good friends. The paper will grow into this book, a book that only scratches the surface of a deeper history.

Kimble D. McCutcheon

Acknowledgements

So many have given of their time and energy in the creation of this book that it is impossible to properly credit them all. However, some have been especially helpful.

Graham White planted the seed with coverage of the Tornado in his book *Allied Piston Aircraft Engines of World War II.* He also freely gave contacts, information, encouragement, and criticism. Dan Whitney provided the original research lead that proved the existence of source material for the Tornado project. Kevin Cameron made gentle suggestions about the craft of writing and technical issues surrounding the unique construction of the Tornado. Without the help of these three, the project would have never been properly started, let alone finished.

Several individuals who either directly or indirectly worked on the Tornado supplied a treasure trove of personal accounts and photographs. Mrs. Edith Abbamont, Mrs. Ruth Daub, Mel DeBlock, John Minnich, Emerson Watson, Dave Way, and Augie Zoll contributed the heart of this story. They fielded scores of telephone calls, reviewed accounts of the engine restoration, and assisted in countless other ways. John Minnich, perhaps the last surviving person who worked on the original Tornado project, told the story of the engine's designer, Rudy Daub.

There were many who assisted with research, reproductions, and photo permissions: Pat Reilly and Jeff Hearne of the Aviation Hall of Fame and Museum of New Jersey; Marilyn Graskowiak, Katherine Igoe, Kristine Kaske, Brian Nicklas, Paul Silberman, and Larry Wilson of the National Air and Space Museum; Julie Wood of the Rolls-Royce Heritage Trust; Mike Speciale and Jack Connors of the New England Air Museum; Garland Gouger and Louise Sykes of the NASA Langley Research Center; and finally, the staff of the National Archives at College Park who go out of their way to provide a most pleasant research experience.

Book design is by Betty Howard of Altherr Howard Design of Huntsville, Alabama. Betty, along with Florence Smith, helped me understand the finer points of book design and production.

I am deeply grateful to the reviewers of the book—Kevin Cameron, Doug Culy, Jim Howell, Graham White, and my wife, Shirley. I am also indebted to the editors—Bobbie Graham and Rachel Mountz . All of these people provided invaluable criticism and corrections. Any remaining faults with the work are entirely mine.

Introduction

The Northrop Alpha was designed in 1929. Its stressed skin construction was to become standard for nearly all aircraft. (Courtesy of Wings on the Web, http://www.wingsontheweb.com)

Prelude

In the decade prior to World War II, U. S. military officers, especially those associated with Army aviation, found themselves operating aircraft that were decidedly second rate. Airliners were faster than their bombers, and privately owned sport planes were faster than their pursuit aircraft. Advances in construction and streamlining held great promise, but missing were powerful engines of low frontal area that would make the next generation of Army medium bombers, attack aircraft, and pursuit ships competitive with those of potential enemies.

The Army initially addressed this problem through internal research and development. However, a completely new engine was beyond the capability of the small staff of Army engineers. The Army began funding commercial engine companies to design and build liquid-cooled engines of between 1,000 and 2,500 hp. Before the end of World War II, no less than seven companies had become involved in the development of at least 13 experimental engines.

In 1940, several exotic engines were being designed to power a new breed of fast aircraft. The most powerful engine of all was the Wright Tornado, a 42-cylinder liquid-cooled radial that had been conceived in 1939 by the Wright Aeronautical Division of Curtiss-Wright. This engine was to deliver no less than 2,350 hp and be no more than 36.5 inches in diameter. It held great promise for being the first piston engine to exceed the 500-mph mark.

Building such an engine was no easy task. It required revolutionary thinking and new design approaches to manage size, weight, and vibration. A dedicated team of brilliant engineers and designers, led by Bill Lundquist and Rudolph Daub, was assisted by equally dedicated and skilled mechanics and test engineers. A 14-cylinder prototype was closely followed by the full-sized 42-cylinder model. These proved that the unusual characteristics of the engine, such as its small diameter, high crankshaft speed, multiplicity of tiny cylinders, and unusual power transfer mechanism, were sound.

Just as the Tornado project entered the extensive testing phase required of all new aircraft engines, the U. S. became involved in World War II. Curtiss-

Wright was overwhelmed with its commitments to produce engines for B-17s, B-25s, and B-29s. Experimental projects like the Tornado languished. In the meantime, air-cooled engines more powerful than the Tornado had evolved, and aircraft designers had learned to install them with little or no drag penalty. The Tornado was eventually abandoned early in 1944, before testing was complete and before it ever flew.

The Tornado project cost the Government over $300,000. It cost Curtiss-Wright over $6,000,000. Only one engine was delivered. Like so many other experiments abandoned after World War II, the Tornado was forgotten and eventually scrapped. It was saved from the smelter by shear luck and finally restored by another dedicated team led by Jerry Abbamont, one of the original engine testers.

Although the Tornado is an evolutionary dead end, it is of great historical interest because it represents one version of the technology that would have been required if gas turbines had been delayed. The quest for higher power, especially from smaller packages, forced engine designers to try radical new concepts. Lessons learned from engines like the Tornado led to better components and engineering practices for jet engines. Today, the Tornado is just a signpost marking the entrance to one of history's many blind alleys. But what an elegant signpost it is, and what a wonderful story it tells of the optimistic experimental engineering, dedication, and craftsmanship of a bygone era.

A word of explanation regarding terminology is in order. Throughout the book, "the Army" has been used to denote the organization in charge of land-based aviation. This is done for the sake of consistency and to spare the reader confusion with terms such as Air Service, Air Corps, Air Forces and the like. Similar liberties have been taken with Wright Aero to designate the Wright Aeronautical Division of Curtiss-Wright, and with Wright to designate the collective of all companies bearing the Wright name from Wright Aeronautical Corporation through Curtiss-Wright.

The Aircraft Engine Industry in the 1930s

Never in the history of aviation has such rapid progress been made as in the decade of the 1930s. The late 1920s designs of Jack Northrop hinted at what was to come. Then, during the 1930s, the all-metal stressed-skin monoplane became commonplace, especially in military applications. It quickly replaced the archaic fabric-covered, wire-braced biplane designs, and became the paradigm for all future aircraft development.

The decade of the 1930s was equally pivotal for development of aircraft engines. Indeed, without these new engines, airplane design would not have moved forward. During the 1920s, the four major U.S. engine makers, Curtiss, Packard, Pratt & Whitney, and Wright Aeronautical, all introduced air-cooled engines. These new air-cooled engines allowed feats in aviation like Colonel Charles Lindbergh's non-stop New York to Paris flight and the economical transport of airmail across the country. Airmail contracts also facilitated the introduction of regularly scheduled passenger service.

The reliability and low weight of the air-cooled engine, coupled with new demand for airline transportation, had aircraft designers scrambling to produce new, fast, efficient designs. The decade of the 1930s would see the introduction of one design still in service today—the Douglas DC-3. All of these new designs used air-cooled engines. Because of installation weight and complexity, the market for liquid-cooled engines was exclusively military, and was primarily for Army fighters. There was little economic incentive for engine makers to use any precious research dollars to develop new liquid-cooled engines.

Despite the rosy outlook for air transportation, military planners, with dreams of high-speed fighter aircraft in their heads, had worried looks on their faces. While air-cooled engines were light, reliable, and easy to service, the most powerful one available at the close of 1929 was the 575 hp Pratt & Whitney Hornet "B." Conversely, no less than nine liquid-cooled engines producing more than 575 hp were available. The most powerful of this lot, the Packard 1A-2775, produced 1,300 hp.[1]

The higher power of the liquid-cooled designs was due in large part to greater specific output. That is, liquid-cooled engines produced more horsepower per cubic inch of displacement than did their air-cooled counterparts. In those early days of engine design, the science of heat transfer was still in its infancy, as was the art of building cylinders with numerous thin cooling fins. Consequently, liquid-cooled engines were more effective in removing heat from the hottest parts of the cylinders than were air-cooled engines. Hot cylinders lead to detonation, which could destroy an engine in seconds.

The aviation fuel used by the military in 1930 was required to be a minimum of 65 octane raised to 87 octane by the addition of no more than 6 ml of tetraethyl lead per gallon[2]. The combination of low-grade fuel and poor cooling made air-cooled engines of the early 1930s unable to use a high degree of supercharging. Liquid-cooled engines could run at much higher manifold pressure without suffering detonation, and therefore, were considerably more powerful than air-cooled engines of similar displacement.[3]

When military planners considered the amount of power that would be required for the next decade's aircraft designs, they realized that no air-cooled engine was even planned that could meet the challenge. Multiple rows of air-cooled cylinders would be required. The mechanical and thermodynamic considerations of such arrangements had just begun to be studied, and indeed would not be perfected until the late 1930s.

Winner of the 1925 Schneider Cup race with a speed of 232.57 mph, the U.S. Curtiss R3C-2 (Courtesy of Wings on the Web, http://www.wingsontheweb.com)

International events had the military planners worried as well. While the rise of expansionist totalitarian governments in Europe pointed to war, the U.S. technology edge seemed to be slipping. One example was rapid progress in aircraft designed for the Schneider Cup competition. The U.S. had won the 1923 and 1925 Schneider Cup seaplane races in Curtiss aircraft under Navy sponsorship. The Cup was lost to the Italians in 1926. Then Great Britain won the cup in 1927 and 1929. The 1929 entry was a Supermarine S.6 powered by an updated Rolls Royce "R" liquid-cooled engine producing 1,920 hp. The winning speed of 329 MPH was a full 96 mph faster than the 1925 U.S. contender.[4] Such an improvement in only four years was astonishing. Great Britain would go on to permanently claim the trophy in 1931 with a Supermarine S.6B powered by the Rolls Royce "R" which by then was producing 2,300 hp and weighing just 1,630 pounds. While the "R" was a pure racing engine, nothing from the U.S. could come close to its power-to-weight ratio.

Winner of the 1926 Schneider Cup race with a speed of 246.496 mph, the Italian Macchi M-39

The Curtiss D-12 provided power for the U.S. Schneider Cup competitors. (Courtesy of the New Jersey Aviation Hall of Fame and Museum)

Three consecutive wins allowed the British to permanently claim the Schneider Cup in 1931 with the Supermarine S.6B. Average speed was 339.82 mph. (Courtesy of The A.J. Jackson Collection)

The Italian team had provided stiff competition for the Schneider Cup, but had also pushed technology too fast. Carburetor problems and airframe development difficulties resulted in the loss of aircraft and pilots. The pursuit of high-speed seaplane records became a matter of national pride. In 1934, the Italians captured the absolute speed record of 440.60 MPH in a Macchi-Castoldi MC-72 seaplane powered by a 24-cylinder Fiat AS-6 producing over 3,000 hp. The record established by this aircraft still stands.

With mounting evidence that the U.S. technological edge was evaporating, Army planners took drastic action to get new liquid-cooled engines—they designed one themselves.

The Rolls-Royce "R" engine powered the S.6B

Holder of the absolute seaplane speed record, the Macchi-Castoldi MC-72

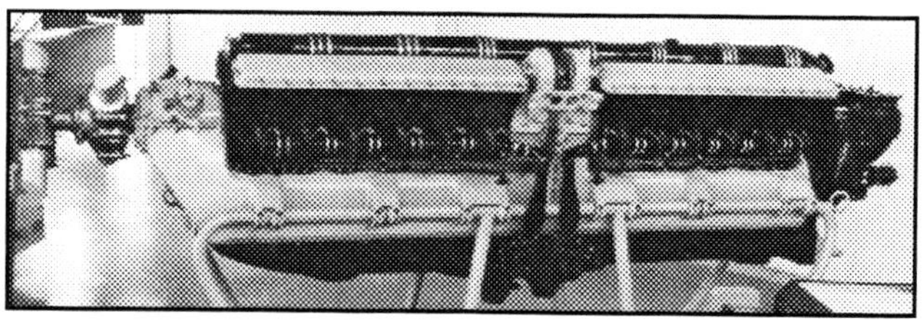

The Fiat AS-6 produced over 3,000 hp

The Army Begins "Hyper" Engine Research

In late 1929, Samuel D. Heron, working at the Army Engineering Laboratories at Wright Field, Ohio, began to experiment with methods of increasing the output of engine cylinders. Heron had worked at the Royal Aircraft Factory in England from 1915 through 1916 with Professor A. H. Gibson. Here the first systematic study of air-cooled cylinder construction was undertaken and the characteristics of modern cylinders were defined. Heron came to the United States in 1921 where he used his considerable knowledge to enhance the air-cooled cylinder.

Around 1929, English engine guru Harry R. Ricardo had claimed that the poppet valve had reached its specific output limit and should be replaced with sleeve valves in high-power designs. Ricardo reasoned that the white hot exhaust valve limited an engine's power output more than the cooler running sleeve valve, and he had test results to prove it. Heron's experiments were intended to investigate these claims. The poppet valves used by Ricardo to draw such a conclusion were not of the internally cooled type Heron had invented. Using a single air-cooled Liberty engine cylinder, Heron fashioned a water jacket around the cylinder and cooled the head with a spray of water. The engine was run at high speed and with high boost pressures. Specific outputs higher than the limit stated by Ricardo were obtained at once. Further experiments were performed using a cylinder with a complete water jacket. These tests yielded results far superior to any engine, liquid-cooled or air-cooled, existing or planned. The Army at once became enthusiastic over the development of a high performance, or "Hyper," engine based on Heron's cylinder design.

Sir Harry Ricardo, Proponent of the Single Sleeve Valve (Courtesy of Ricardo Consulting Engineers Ltd.)

The Hyper cylinder was to have the same 4.625-inch bore as the Liberty engine cylinder used in Heron's experiments, but the stroke was to be shortened to 5.000 inches to allow higher speeds of rotation. This gave a displacement of 84 cubic inches per cylinder. The engine also featured a hemispherical combustion chamber with two valves and two spark plugs.

Also in vogue at the time was the concept of high-temperature cooling. The Army thought that by using pressurized ethylene glycol at a temperature of 300° F, radiator size could be reduced to a point that liquid-cooled engines would be competitive with air-cooled ones on a weight basis. Cylinders of such high output were more easily cooled if constructed individually rather than cast en bloc. The Army insisted on separate cylinders, despite difficulties with leaks, engine length, and engine stiffness. The Army rationalized that separate cylinder construction made possible refinements in the cylinder barrel, head, and jacket that resulted in better cooling and relieved stresses at maximum loads. Ease of service and repair was also a consideration.[5]

The Army placed an additional and undue emphasis on engine frontal area. Army engineers demanded engines that could be submerged in the wings of bombers, and streamlined into the fuselages of fighters. This resulted in opposed-cylinder configurations.

By 1932, the Army had conceptualized the design of a Hyper engine. It was a 12-cylinder horizontally-opposed separate cylinder engine of extremely shallow depth, using the Hyper cylinder, and a coolant temperature of 300° F. This engine was to have a takeoff power of around 1,000 hp.

On October 5, 1932, the first purchase order was issued to the Continental Motor Company for engineering development of the Hyper engine.[6] Continental was a successful manufacturer of automobile and truck engines that had entered the aviation market with an air-cooled radial in 1928. Coincidentally, Continental owned world rights for the single sleeve valve (the type advocated by H. R. Ricardo), and already had a contract with the Navy for a radial air-cooled sleeve valve engine.

Samuel D. Heron, Developer of the Hyper Cylinder (Courtesy of Propulsion Directorate, Air Force Research Laboratory)

Precisely how the Army decided to contract with the very company who owned rights for the competing technology and that had such an obvious conflict of interest is an interesting question. Continental historian William Wagner maintains that the Navy contract helped Continental get the Army development contract. A special office was set up in Dayton*, Ohio, under the directon of Harold Morehouse. According to Continental engineer Carl Blanch, "There was a race on, and I was the conductor of the tests on both the single-sleeve in one room, and the hyper-cycle in the other. We got to the point the single-sleeve performed the highest horsepower per cubic inch of anything of that era. Following that—six months later—we made the Hyper do the same thing."[7]

The Continental O-1430, First of the Hyper Engines (National Archives and Records Administration)

It may be that the Army's choice of Continental for the actual development was more a matter of who they could control. While Continental with its sleeve-valve had a serious conflict of interest and much less expertise in the development of high power than others, it was in such poor financial shape that it could not afford to ignore any new business. The largest engine manufacturers, Pratt & Whitney Aircraft and Wright Aeronautical Corporation, were busy with air-cooled development and wanted no part of a liquid-cooled engine over which they had little or no engineering control.

Continental's lazy development pace fueled the controversy. The Army, having designed the cylinder and basic engine layout, limited Continental's role to routine engineering and testing. Yet, it still took six years to produce a complete engine. Part of the problem was the

*William Wagner places the Hyper development in Detroit, while Robert Schlaifer says Dayton. Since the customer was near Dayton, it is probable that the cylinder was developed there.

Army's insisting on 300° F coolant temperature and the 12-cylinder opposed-cylinder configuration. But Continental, either from inability to fund any of the development themselves, or from lack of interest in the project, insisted on a signed contract for every expense, no matter how small.[8] Further complicating matters was the exceedingly low level of funding available for engine development.[9]

In the end, Army dropped the 300° F coolant requirement. Experience showed that with the high coolant temperature, so much heat was transferred to the oil that the increased oil cooler size offset the decreased coolant radiator size.[10] Army also dropped the opposed cylinder configuration in favor of an inverted vee. The concept of engines of extremely shallow depth for submerged installation also fell from favor when airframe manufacturers found that buried engines produced insurmountable structural, fuel placement, landing gear placement, and engine access problems.

Continental ultimately refined the O-1430 into the IV-1430, which, given time and money for development, would have probably become an acceptable engine. However, World War II intervened, and Continental instead became a supplier of Rolls Royce Merlins and Pratt & Whitney Wasps under license.

The Continental IV-1430, Final Evolution of the O-1430 (National Archives and Records Administration)

In the meantime, two things had happened to change the development picture. First, the aircraft engine industry had finally felt the effects of the Great Depression. Airline orders had gone flat, as had the market for military air-cooled engines. This made engine manufacturers more disposed to pursuing liquid-cooled development than they had been in 1932. Second, Army became so disillusioned with the pace of Continental's engine effort that it was willing to look elsewhere for liquid-cooled development. In the late 1930s and early 1940s, similar development contracts were placed with Allison, Lycoming, Chrysler, Pratt & Whitney, and Curtiss-Wright. At least 12 different engines were built and run because of the Army's Hyper program. It is the Curtiss-Wright side of the Hyper story that concerns us here.

A Short History of Liquid-Cooled Engines at Curtiss-Wright

Liquid-cooled engines built by the Curtiss Aeroplane and Motor Company were, until the end of 1926, better than the best Europe could offer. The Curtiss D-12, which had appeared in 1922, won the 1923 and 1925 Schneider Cup races for the United States. The engine was so successful that the British Air Ministry awarded a contract to Rolls Royce for development of an engine along the same general lines. The Rolls-Royce "Kestrel" resulted from this activity.[11] Had the D-12 been installed in a more modern airframe, it would have probably won the Schneider Cup for the United States in 1926.

The Curtiss "Conqueror," a 1,550 cu in, 575-hp successor to the D-12, was put into production in 1926. Concurrent with this was the introduction of the first high-power, air-cooled engine, the 400 hp Pratt & Whitney "Wasp." More air-cooled engines of still higher power were soon to follow, from both Pratt & Whitney and Wright. Because of these improved air-cooled engines, the market for liquid-cooled engines disappeared overnight. The Navy committed to air-cooled engines in all applications except those requiring the highest power. All commercial engines were air-cooled. The only exception was engines for Army pursuit aircraft. Since most of these were coming from Curtiss, Packard stopped building liquid-cooled engines to concentrate on its air-cooled diesel intended for commercial applications.

The Curtiss Aeroplane and Motor Company merged with Wright Aeronautical Corporation in 1929 to become Curtiss-Wright. Initially, both Curtiss and Wright engine organizations continued to operate as separate companies. In 1931, the groups were combined as Wright Aeronautical Corporation (hereafter referred to as Wright Aero). During this time, the Conqueror continued to improve, albeit slowly. Horsepower increased from 575 to 650. However, when Wright Aero introduced the GR-1820 Cyclone "E" in 1930, the corresponding sales explosion absorbed all of Wright Aero's engineering talent. By 1932, further development of the Conqueror had ceased.

Robert Schlaifer sums up the military situation splendidly:

> "It was clear by 1932 that if the development of liquid cooled engines was left to the same financing and the same incentives as the development of air-cooled engines, the liquid cooled engine would soon be hopelessly obsolete. Already by 1932 American liquid-cooled engines, which had been as good as any in 1925, were inferior to the Rolls-Royce Kestrel. But although it was highly probable that the air-cooled engine would continue supreme in most applications, it was by no means possible at any time in the 1930s to demonstrate conclusively that this would always be true of all applications. Insofar as aircraft engines were put to civilian uses, all these possibilities could be rightly neglected; the "expectation" of profits from liquid-cooled development was far too small to justify the large sums which it would have to cost if success was to be even possible, and in a rational economy an unprofitable undertaking is a wasteful one. For military use, however, the stakes were higher. If the liquid-cooled engine proved ultimately to give superior performance, even if it should be only in a single important use of a single type of aircraft, an air force without a good liquid-cooled engine would be at a most serious disadvantage in case of war. And if such a superiority should be demonstrated by foreign liquid-cooled engines after development of the type had been neglected for a number of years in this country, it would require a long and perhaps fatal period to overcome the disadvantage."
>
> "Thus there can be no question that the services were fully justified in doing all they could to foster the development of liquid-cooled engines. Every major foreign country consistently promoted the development of both air-cooled and liquid cooled engines, no matter which type seemed the better at a given moment. The two most successful countries, Britain and Germany, not only supported development of both types, but maintained competition in each field by having two major firms working continually on each type. The abandonment by the United States of either type would have been a completely indefensible gamble."[12]

The Army was convinced it had to pursue development of liquid-cooled engines, but also was convinced that Wright Aero was not interested in the project.

Continued support for liquid-cooled engine development at Wright Aero did come from the Navy during the first half of the 1930s. Navy was concerned with the growing superiority of foreign airplanes as demonstrated in the 1929 and 1931 Schneider Cup races. Consequently, Navy embarked on a special "high-speed development program" in 1930. While this program was hopelessly underfunded, it did produce two engines.

The first began life as the H-1640 "Chieftain," a 12-cylinder, double row, air-cooled radial of extremely small diameter. Curtiss designed this engine in 1926 at a time when undue emphasis was placed on frontal area and, therefore, engine diameter. Although this engine was unsuccessful, Curtiss and Navy decided in 1930 to dust off the basic design, to replace the cylinders with slightly larger liquid-cooled ones, and ultimately to expand the engine to four rows of six cylinders each for a displacement of 4,240 cu in. This new engine was to develop 2,000 hp.

Development was begun on a two-row portion which was called the H-2120*.The small diameter resulted in excessive connecting rod angularity as well as peculiar motion for all but the master rods. A great deal of effort went into the perfection of a "true-motion" connecting rod arrangement that was ultimately abandoned. Work continued from 1933 until 1936 when a 100-hour development test was run at 1,000 hp. Then the Navy changed its mind and reverted to its policy of using and developing air-cooled engines exclusively. Wright Aero quickly dropped the project, convinced that the design was mechanically unsound.[13]

The Wright Aero H-2120, an early and unsuccessful small-diameter liquid-cooled radial (Courtesy of the New Jersey Aviation Hall of Fame and Museum)

*The "H" designation for a radial engine is unique to the "Chieftain" line of radials. It presumably stands for "hex."

The Navy also supported Wright Aero in the development of a second liquid-cooled engine from 1932 until 1934. This engine, the V-1800 was to replace the Conqueror. It was a 12-cylinder vee with individual cylinders specifically designed for high-temperature cooling. It was also to be supercharged to maintain its sea-level power rating to an altitude of 12,000 feet. The engine ultimately ran a 50-hour development test at 800 hp in June of 1934. Shortly thereafter, the Navy had to abandon the project due to lack of funds. The engine was sold to Russia, and enough work done at Russian expense to increase the power to 900 hp.[14]

The last 100 Curtiss Conquerors were sold to the Army in 1935, bringing to an end V-12 liquid-cooled engine development at Wright Aero.

Wright Aeronautical Reenters Liquid-Cooled Development

About 1938, when Continental's lack of progress on the O-1430 became intolerable, the Army gave Wright Aero a contract for the development of a 24-cylinder, liquid-cooled engine suitable for submergence in an airplane wing. Originally, the Army intended to install the engine in fighters. When it became clear the wings of fighters were too thin, the Army turned its attention to installing the engine in the wings of attack aircraft.[15]

Wright Aero initiated a design study in September of 1938 to determine the liquid-cooled engine layout that would be most suitable for submergence in the relatively small wing used on attack bomber aircraft. The engine was to have an initial rating of 1,500 hp and was to grow to 1,800 hp. For the studies, designers chose initial values of 200 psi for the brake mean effective pressure (BMEP) and 3,200 feet per minute (fpm) for the piston speed. Rough designs were made of four engine configurations in November 1938 and an analysis was prepared. The results of this activity are summarized in Tables A.1 through A.10 of Appendix A.

From this analysis, it was decided that the Flat X configuration was the most feasible design because it was compact and most easily submerged. In discussions with airplane manufacturers, it became apparent that an initial takeoff power rating of 1,800 hp was desirable to meet aircraft performance requirements. Accordingly, displacement was increased to 2,185 cu in. Wright Aero then undertook extensive detailed component design and produced an installation drawing that was released to aircraft manufacturers in January 1939.

Numerous and severe structural and installation problems were identified with the buried installation. The principal ones included:

- Interference with the landing gear
- Interference with the front spar
- Front spar design difficulties when accommodating engine removal
- Engine width interfered with desirable close bulkhead and rib spacing
- Length of engine required undesirably wide spar spacing in small wings
- Engine location reduced available wing space for fuel and other components
- Removable wing skin panels reduced wing strength, increased weight, and required additional bracing
- Difficult engine access
- In small thin wings, engine thickness (21.0 inches) required undesirable bumps outside the normal wing contour
- Installation in a monospar wing was practically impossible

In order to address some of these difficulties, the external arrangement of the engine was revised to separate the turbosupercharger and intercooler from the engine, and an updated installation drawing was circulated. Aircraft designers continued trying to squeeze this engine arrangement into their airframes. As both aircraft and engine design progressed, it became increasingly apparent that the Flat X was not as practical as was originally believed.

In April of 1939, Wright Aero started preliminary design of a small diameter liquid-cooled radial engine that would give the same performance as the Flat X. The following assumptions were made:

- Engine installation in a conventional nacelle is most practical from a wing structure and engine accessibility standpoint.
- Drag of an airplane equipped with nacelles will compare favorably with drag of an airplane equipped with submerged engines if the nacelle diameter is sufficiently small and the wing structure weight penalty for submerged installation is considered.
- The radial engine involves fewer basic design problems than the Flat X.
- The radial engine is universally applicable to all existing aircraft types.
- The outside diameter of the radial engine would be acceptable if it were slightly larger than the 21-inch nose extension of the Flat X.

Several rough radial engine configurations were studied and a 42-cylinder engine with a displacement of 1,965 cu in emerged as the most suitable. This engine was to have a bore of 4.125 inches and a stroke of 3.500 inches. It consisted of six rows of seven cylinders per row with an outside diameter of 33.0 inches. This diameter allowed arrangements for engine mounts, exhaust manifolds, and accessories. The design was analyzed and found to have the following desirable features:

- Rigid crankcase
- Small cylinder bore
- Low BMEP (169 at 1,800 hp with a 2,500 fpm piston speed)
- Smooth operation
- Output can be increased to 2,400 hp with 56 cylinders
- Reduced engine weight compared to Flat-X submerged installation
- Basic design will cover a power range of 300 to 3,000 hp with multiple cylinder banks
- High horsepower per square foot of frontal area (303 hp/sq ft at 1,800 hp)
- One piece master rods
- Accessibility when installed in nacelle
- Efficiently applicable to all airplane types
- Adaptable to air-cooling

The following undesirable features were also identified:

- Multiplicity of cylinders
- Length
- Complicated ignition system
- Complicated fuel metering
- High rpm (4,290 rpm for 2,500 fpm piston speed)

Allowances were made in the basic design to allow for air cooling in the future, with a corresponding diameter increase to 40.0 inches. It was acknowledged that forced-air cooling would be required due to the small cross sectional area of the engine and the volume of air involved. Studies were made of five, six, and seven cylinders per row. The seven cylinder per row configuration had the most attractive combination of firing order, bore/stroke ratio, cylinder size, and overall diameter. Configurations with nine cylinders per row were not seriously considered because they left insufficient room between cylinders for manifolds and ignition leads.

This artist's concept of the Wright R-2160 "Tornado" was used as a marketing tool before photographs of actual engines were available. (Arthur Nutt Papers, Acc. 1987-0015, National Air and Space Museum, Smithsonian Institution, SI Neg. No. 2001-833)

Attention was given to smaller and larger engines that used multiples of the basic row design.

Early in June 1939, a new installation drawing describing a 1,800 hp engine was circulated to aircraft manufacturers. Though very enthusiastic about the engine, they felt it would be even better at 2,500 hp. To comply with this demand for more power, the bore was increased to 4.250 inches and the stroke to 3.625 inches, resulting in a displacement of 2,160 cu in. The initial BMEP was raised to 210. This change yielded 2,350 hp at 4,145 rpm, with a piston speed of 2,500 fpm. To shorten the engine as much as possible, the air-cooling option was eliminated and the valve gear revised. A substantial 3.5-inch length reduction to 95.5 inches was achieved at a cost of 1 inch in diameter.

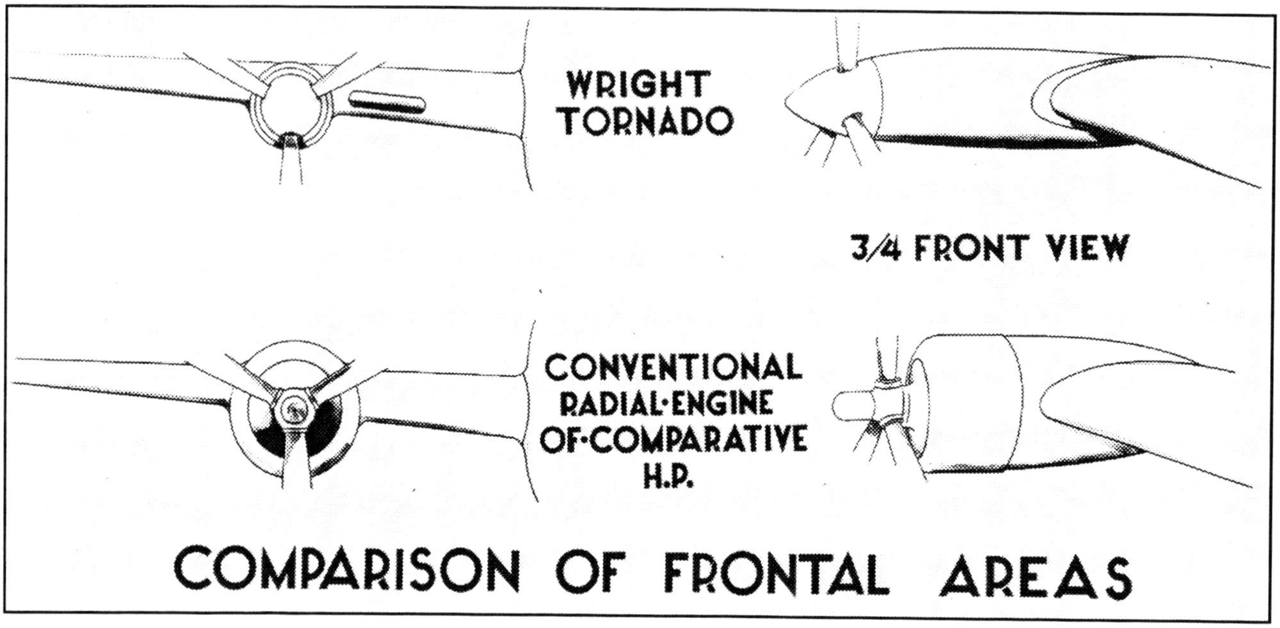

Another marketing angle exploited the extremely small diameter of the Tornado. (Arthur Nutt Papers, Acc. 1987-0015, National Air and Space Museum, Smithsonian Institution, SI Neg. No.2001—832)

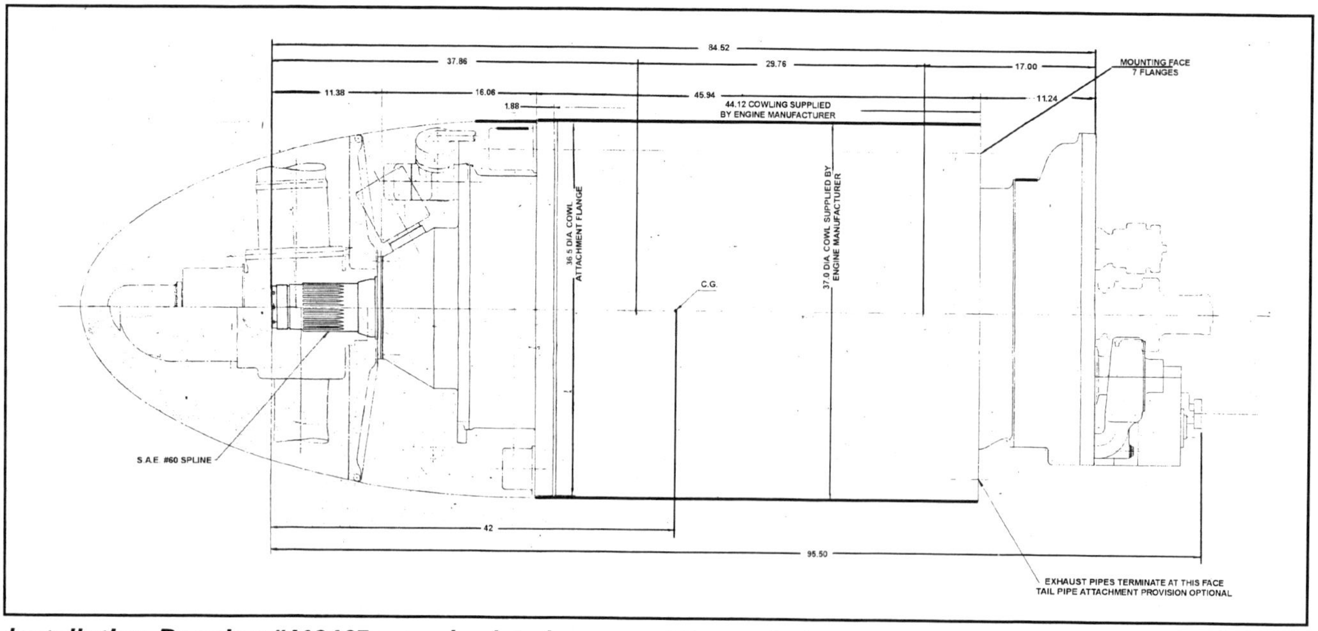

Installation Drawing #413465 was circulated among airframe designers to garner support for the Tornado. (From National Air and Space Museum, Technical File BW-730120-01, Smithsonian Institution)

The overall diameter continued to grow as detailed design progressed until a final value of 36.5 inches was reached. The target weight was 2,400 pounds. The coolant and oil return lines were routed through an external 2-inch by 10-inch blister in the nacelle directly below the engine. Finally, a single speed, single stage supercharger was specified. This configuration was detailed in Installation Drawing #413465.[16] Wright Aero, in keeping with its tradition of naming engines after windstorms, had selected an appropriate name for its most powerful engine yet. The "Whirlwind" that had grown into a "Cyclone" was now a "Tornado."

The Tornado is Unveiled

Wright Aero introduced the Tornado concept in its *Summary Report of Flat and Radial Liquid Cooled Engine Design Studies*, released on August 30, 1939. It was the *Supplement No. 1* to the report, entitled "Tornado Engine Model Variations," that first unveiled the concept in detail. These proposals were submitted to the Army in hopes of winning a development contract. The following discussion explores some of the unique features.

The general design of the engine was a composite of the desirable features of both radial and in-line configurations. The cylinders were cast as a one-piece bank that was mounted lengthwise on the engine, with seven banks forming the complete engine. Forged steel sections were bolted together to form the crankcase. The basic crankcase module was a 2-row 14-cylinder unit. Engines of any arbitrary size from 14 to 70 cylinders could be constructed by bolting together multiples of the basic module and by casting cylinder banks of 2, 4, 6, 8, or 10 cylinders per bank. One intake and one exhaust valve per cylinder were operated by small rocker levers driven by a single overhead camshaft per bank.

The built-up crankshaft was constructed of multiple two-throw segments, one per module. Power from each crankshaft module was transmitted to the reduction gear by seven high-speed layshafts situated between the banks of cylinders. In addition to breaking the overall vibration problem into manageable 14-cylinder chunks, this crankshaft arrangement had the following advantages:

All of the highly loaded connecting rod and main bearings were solid rather than split. Experience with high-powered engines had shown split bearings to always be inferior to ones of solid construction.

Gross torque on individual crankshaft throws was reduced to the maximum torque for a 14-cylinder module. This allowed the use of small crankpin and main journal diameters compared to engines with longer crankshafts where gross engine torque must pass through each succeeding crankpin and main journal on the way to the reduction gear. The reduced crankpin diameter reduced master rod big end weight, allowing higher operating rpm and lighter crankshaft counterweights.

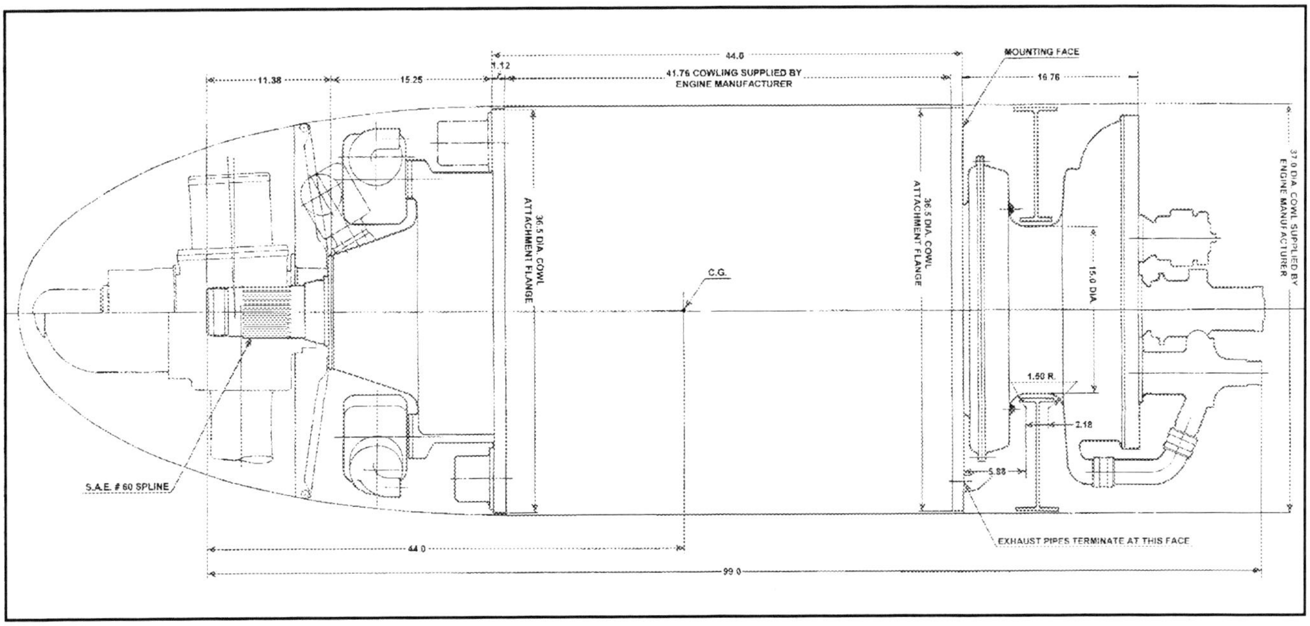

Installation Drawing SU-333 presents one solution to structural problems with submerged engines. (From National Air and Space Museum, Technical File BW-730120-01, Smithsonian Institution)

Engine output could be varied in units of 14 cylinders, allowing proportional power outputs. In the larger engines, rpm did not suffer from the handicap of proportionally larger crankshaft diameter and connecting rod big end weights.

The range of power output expected of the basic Tornado design is detailed in Table 1-1.

The initial takeoff power ratings proposed by Wright were based on experience with large-bore (6.125 inches) air-cooled engines with piston velocities and BMEPs in excess of the proposed values. It was generally accepted that cylinders with smaller bores are more adaptable to operation at high BMEP. The 69 percent reduction in bore (from 6.125 inches to 4.250 inches) was anticipated to help make the planned power ratings possible while still retaining expected reliability.

Since it was believed that the Tornado would be used in a variety of high-performance aircraft, Wright designed the engine to accept a number of different propellers. These included left and right rotating traditional propellers, coaxial contra-rotating propellers, and provisions for a two-speed propeller reduction gear. The particulars for these variations are summarized in Table 1-2.

Additional length for the contra-rotating propeller drive was due almost entirely to provisions for the second propeller shaft. All of the propeller drive arrangements were to be equipped with a hydraulic torquemeter, the inclusion of which accounted for about 50 pounds of additional weight. All of the proposed propeller drive variations were accomplished within the nose section of the engine so that no change in crankshaft rotation or accessory drives was required to achieve a change in propeller rotation.

Table 1-1. Specifics for Various Tornado Variations

Number of Cylinders	14	28	42	56	70
Displacement (cu in)	720	1,440	2,160	2,880	3,600
Takeoff Rating (bhp)	780	1,577	2,365	3,160	3,945
Est Weight (lbs.)	888	1,680	2,400	3,140	3,800
Diameter (inches)	36.5	36.5	36.5	37.5	38.5

Table 1-2. Tornado Propeller Drive Variations

Propeller Type	Gear Ratio Range	Length Increase	Weight Increase
Clockwise	3:1 to 6:1	none	none
Anti-clockwise	3:1 to 6:1	none	none
Contra-rotating	3:1 to 6:1	11"	150 lbs
Two-speed	30% increase	6 to10"	150 lbs

Supercharger Arrangements

The Tornado engine design allowed convenient application of any existing type of supercharging. While the first experimental engine was to be equipped with only a single-stage, single-speed internal supercharger, other arrangements were contemplated to meet future demands.

Table 1-3 briefly describes the performance, weight, and engine overall length increase of the Tornado with the contemplated turbosupercharger; two-speed, single-stage supercharger; and two-stage supercharger options. These weights were based on the use of a single rotation propeller without a torquemeter, and did not include intercooler or ducting weight.

No engine outside diameter increase was required with any of the specified supercharger arrangements.

Fuel Injection System

Accurate metering and precise distribution of the fuel-air mixture is a critical problem in any multi-cylinder engine. This problem was accentuated considerably in the Tornado design due to the large number of cylinders. Wright Aero, cognizant of these challenges from the start, produced a fuel injection design that was both unconventional and promising.

Starting with a conventional Stromberg PD pressure carburetor, Wright Aero proposed rearranging the components to suit the special needs of the Tornado, reducing the bulk of the fuel metering system and providing a means of controlling the mixture distribution to each cylinder.

Airflow to the engine was measured in the conventional way by impact tubes and a boost venturi at the supercharger entrance. The pressure difference obtained at the venturi was transferred via two channels to seven small fuel flow regulators located at the start of each of the seven intake manifolds. Fuel from the engine driven fuel pump was supplied to the seven regulators where it was metered in proportion to the airflow. At this point, the regulated fuel entered a series of channels within each individual intake manifold. Wright Aero believed that by varying the number and length of channels, it would be possible to have total control of mixture strength at each cylinder.

While the fuel system was complex, it fundamentally consisted of seven identical six-cylinder in-line fuel injection units. This reduced mixture distribution problems, resulted in physically small components that contributed to a small engine diameter, and enhanced engine reliability by duplicating critical components.

Table 1-3. Tornado Supercharger Variations

TURBOSUPERCHARGER

Ratings	bhp	Altitude, ft
Takeoff	2,350	Sea level
Military	2,350	5,000
Normal	2,000	18,000

Weight – Dry – 2,400 lbs. without intercooler

Length (overall) – 95.5 inches

TWO-SPEED SUPERCHARGER

Ratings	bhp	Altitude, ft
Takeoff, Low Blower	2,350	Sea Level
Military, Low Blower	2,350	2,500
Normal, Low Blower	2,000	4,500
Military, High Blower	1,975	16,000
Normal, High Blower	1,800	16,000

Weight – Dry – 2,450 lbs.

Weight increase compared to turbo – 50 lbs.

Length increase compared to turbo – 3 in.

TWO-STAGE SUPERCHARGER

Ratings	bhp	Altitude, ft
Takeoff	2,350	Sea Level
Military	2,350	2,500
Normal	2,000	4,500
Military, First Stage	2,180	9,300
Normal, First Stage	1,850	9,600
Military, Second Stage	2,025	20,000
Normal, Second Stage	1,725	18,200

Weight – Dry – 2,575 lbs. without intercoolers.

Weight increase compared to turbo – +175 lbs.

Length increase compared to turbo – +12 in.

Provisions also existed for direct fuel injection. The cylinder heads provided space and access for a fuel injection nozzle in each combustion chamber. Bosses for these nozzles were located on the intake side of the block next to the intake port. Seven drives for small six-cylinder fuel injector pumps were located at the front end of the camshaft on each cylinder bank. The injector nozzles were to be connected to their respective pump units by means of external lines. Control of fuel-air ratio was a problem left for solution at a future date.

At the time, the Army favored direct fuel injection, but no one had yet produced a successful version. Since wild promises were already flying, Wright Aero also threw in the ability to use safety fuel, a low-volatility fuel that was also a hot topic with the Army at that time. Even wilder was a proposed design change to the supercharger drive housing. This change introduced a small diameter point at the rear of the engine intended to facilitate passing the wing spar around the engine in certain applications. Detailed in drawing SU-333, this change was thought desirable for a single engine pusher being proposed (the Vultee XP-54 "Swoose Goose?"). The change involved a six-inch length increase and a weight increase of about 25 pounds.

Dynamic suspension was studied and found applicable to the engine design; however, Wright Aero did not feel it would be necessary. The large number of small cylinders was expected to yield an engine that was considerably smoother than the big-bore air-cooled radials then available. Proof was postponed until an engine was built.

The Army Orders the Tornado

A contract was given in June of 1939 for an experimental two-row section of the R-2160.[17] The two-row section became known as the Tornado 14 or T-14.

Now that Wright Aero had the interest of the Army, it had to turn its dream design into working hardware. Development of high-powered multi-cylinder radial engines is plagued with numerous difficult problems. The engineers, designers, and technicians at Wright Aero addressed these problems in unique ways. The following chapters introduce some of the people behind the engine and provide insight into how the problems were solved.

---

1 Kneen, *Everyman's Book of Flying*, 282.

2 McFarland, "Higher, Faster and Farther," 105.

3 Schlaifer, *The Development of Aircraft Engines*, 251.

4 Pecastaingts, "Schneider Trophy."

5 "Continental's I-1430".

6 *Case History of the IV-1430 (Continental) Engine* - Summary, 1.

7 Wagner, *Continental! Its Motors and Is People*, 73.

8 Ibid. 74.

9 Schlaifer, 271-272.

10 Schlaifer, 269, Note 4.

11 Schlaifer, 247.

12 Schlaifer, 262-263.

13 Schlaifer, 266.

14 Schlaifer, 267.

15 Schlaifer, 287.

16 Goodman, *Flat and Radial Design Studies.*

17 Schlaifer, 287.

2 Process, People, and Ideas

A very young Rudolph Daub at work in his native Germany. After immigrating to the United States, Daub played a significant role in the design of a number of Wright Aero engines. (Courtesy of Ruth Daub)

As with any large project, numerous people contributed ideas and talent to the Tornado. However, one designer and two engineers were the "brain trust" behind the engine. This chapter briefly describes the process by which a new aircraft engine is created, and relates what is known about the lives and contributions of the Tornado's creators.

The Creative Process—from Promises to Performance

As has been seen in Chapter 1, a sort of negotiation took place between Wright Aero, the Army, and various airframe manufacturers while the Tornado was being marketed and sold. At first, the requirements for the engine were theoretical and amorphous. As discussions were held with all the players, the size, shape, and power of the engine changed. In an iterative process, murky initial concepts gave way to concrete dimensions, weights, mounting schemes, and performance guarantees. Eventually, everyone was more or less happy with the result. It is rare that such a negotiation pleases all parties. The best that can be hoped for is an equitable distribution of dissatisfaction.

During this early negotiation phase, teams worked feverishly behind the scenes. At Wright Aero, designers and engineers originated concepts to solve specific problems identified by the Army or by the airframe manufacturers, and made initial plans for how these concepts would be implemented. They produced trade studies showing impact on power, size, weight, and cost for each change to the original concept. Aeronautical and structural engineers at the airframe companies reviewed each new engine proposal to determine aircraft performance and cost impacts. The Army oversaw the entire process, making sure that the final airframe/engine combination satisfied its mission requirements at an affordable cost.

Once specifications for a new engine were finalized, it was the job of the chief designer and chief engineer to turn the rough initial concepts into a product. Often, in the heat of the negotiation, things were promised that would be difficult, if not impossible, to achieve. Good designers and engineers made allowances for this in their weight and power calculations, but there were always challenges to be overcome. Some were known from the outset, but others were nasty surprises that could have never been predicted. Still, all the challenges had to be worked around.

The job of the chief designer was to originate the basic concepts and to harmoniously combine them into a whole. Good design goes beyond simply meeting specification – it is a graceful and refined artistic expression. The ability to visualize a complex aircraft engine with its thousands of parts and to depict that visualization on paper is a gift. The designer must be able to think in pictures instead of words and must be able to communicate with others via standardized drawing formats.

The chief designer would begin by drawing a full-sized layout of the complete engine, making initial assumptions about the size and placement of parts. Scores of detail designers would then produce working drawings for each part. The chief engineer and his staff made sure that each part functioned correctly and that the highly stressed ones were strong enough. The chief engineer also kept a weather eye on the engine weight budget, overall size and shape, and final performance goals. Again, this process was iterative, but finally yielded a set of drawings from which parts could be built.

Individual parts were typically constructed at Wright Aero in the experimental machine shop, or were contracted to outside vendors. Once built, parts were assembled into subassemblies and finally a complete engine by skilled experimental mechanics. Since an experimental engine had no assembly manual, these mechanics had to rely on experience and rough notes and sketches from the designers when the engine was being built.

Occasionally, it was found that a part would not fit. At some point in the complex chain of events, a dimension had been miscalculated or an interference with some other part inadvertently overlooked. In some cases, the experimental mechanic could modify the part or its surrounding parts to make it fit, and modify the drawing so that future production parts would fit correctly. In most cases, however, it was back to the drawing board to redesign the part and remove the error.

Once the engine was completely assembled, it became the responsibility of test engineers. The test engineers were typically young and just out of college with a degree in engineering but little or no experience. They were charged with the task of installing the engine in a test cell, supervising the running of the engine according to a predetermined test schedule, and collecting data associated with the test. If there were problems, they might also offer an opinion as to how they could be solved. Typically, a test session involved three people: the test engineer, an assistant who helped gather data, and an engine operator who manipulated the controls as the test procedure dictated. Some engine runs were brief, others might go on continuously for hundreds of hours and involve three shifts of test engineers, assistants, and operators.

The idea of these long endurance runs was to operate the engine progressively harder until something broke. The broken part was then analyzed, redesigned to eliminate the offending flaw, and the whole process repeated. Given sufficient time and money, an aircraft engine that was light, powerful, and reliable emerged from the process. Hopefully, it would even meet the original specification and please the customer.

Rudy Daub—Chief Designer

Rudolph (Rudy) Daub was the chief designer of the Tornado[1]. If the Tornado were human, it would have the precise, elegant, and complex personality of Rudy Daub. Born on March 15, 1901 and educated in Germany, Daub immigrated to the United States in 1926 and went to work as a draftsman at Wright Aeronautical Corporation in 1928. He became a U.S. citizen on December 30, 1932.[2]

A gifted designer, Daub soon began to produce work that was patentable. He also began to challenge the pecking order within the conservative Wright Aero organization, something that would eventually cost him dearly. At that time, Wright Aeronautical Corporation had no patent department. Instead, all patents went through one man named Roland Chilton. Chilton was the only man at Wright Aeronautical Corporation that had not signed his patent rights over to the company. Instead, he was paid royalties for the use of his patents. It was reported during one year of World War II that Chilton, because of his patent royalties, was one of the highest paid men in the U.S

Rudolph Daub, Tornado, Chief Designer (Courtesy of Ruth Daub)

Daub found that a patent sent to Chilton would often be returned to its designer with a note stating it had no merit. Then, two or three months later, a similar patent would be filed under Chilton's name. By that time, Daub had become a good friend of Bill Lundquist, a rising engineer at Wright Aero. The two of them formed a holding company for their patents. When Wright Aeronautical Corporation management was informed of this, both Daub and Lundquist were fired on the spot. After they told their side of the story, both were rehired that same afternoon, and a patent department was established at Wright Aeronautical Corporation under the direction of Mr. Ken Lane.

In 1929, Wright Aeronautical Corporation merged with Curtiss Airplane and Motor Company. With Curtiss came famous engine experts such as Dr. Arthur Nutt. Wright Aeronautical Corporation, on the other hand, was still smarting from the 1925 gutting of technical staff when Frederick Rentschler formed Pratt & Whitney Aircraft. Wright Aeronautical Corporation had limped along with borrowed talent from the Massachusetts Institute of Technology and the Army's engineering laboratory at Wright Field, Ohio. After the merger, the old Wright Aeronautical Corporation staff was shoved to the side to make room for the bright boys from Curtiss. But with a few exceptions, the Curtiss designers did not impress Daub. He quickly surmised that their act was all flash and no substance, telling coworkers, "I would not let them design a wheelbarrow for me!"

Daub did not kindly tolerate fools in the workplace. In typical Teutonic fashion, he never hesitated to speak his mind, no mater how impolitic. This same frankness extended to other matters. When Franklin Delano Roosevelt started to push his "New Deal," Daub complained loudly and often, acerbically referring to the President as "that SOB." Daub hated organized religion as well. He developed a reputation in some circles at Wright Aero as the sort of troublemaker that one would not necessarily want as a friend. At home, he was warm and affectionate, a good provider who was adored by his wife and son. He loved the music of Richard Wagner and often attended the Metropolitan Opera in New York City.[3]

The R-2600

The Boeing Aeroplane Company approached Wright Aero in 1935 about an engine of 1,500 hp suitable for use in their Model 314 "Yankee Clippers." Raymond W. Young, then director of engineering, organized a design competition for the engine among the 12 top designers at Wright Aero. Daub, at that time, was still "just a draftsman." His friend, Bill Lundquist, suggested that Daub make preliminary drawings for a 14-cylinder, 2-row engine design to be presented along with the other 12.

On Thanksgiving Day of 1935, the 12 assembled at Young's house to submit their designs. Lundquist showed up with the unsolicited one from Daub. The next morning, Daub was called into Young's office and informed that his design had been selected for production, and that he was to be the chief designer of the R-2600 "Cyclone 14." While this put Daub on the map as an engine designer, it did not win him any friends within the Wright Aero design organization. Daub insisted that only Young sign drawings for the R-2600. This action sidestepped the usual collection of signatures that are required on drawings, cutting the political red tape and assuring that Daub's ideas were really implemented instead of being stalled by his growing collection of detractors.

The R-2600 program was kicked off on December 1, 1935. Eleven months later, the engine passed its 150-hour endurance test with a total time on the engine of only 175 hours! Intensely proud of the R-2600, Daub refused to leave the test cell until the test was complete. He wanted to make sure it performed flawlessly.

Daub was in large part responsible for getting the engine together so rapidly. Using proven component designs, Daub created just enough new parts to get the job done. One exceptional innovation was inclusion of the bores for valve tappets in the crankcase casting, rather than in the nose case as had been the previous practice. Daub wanted to use a forged steel crankcase of his invention but was prevented from doing so by the conservative Wright Aero management. Later, his crankcase design outlined in Patent No. 2,189,600 became a standard part of all Wright Aero engines. Daub's ideas greatly reduced the number of separate components and complex machining operations necessary to produce engines.

R-3350

Daub was slated to also design the R-3350 but was assigned the Tornado project before detailed design of the R-3350 could be completed. The R-3350 design was handed over to other designers who, jealous of Daub, removed features he had included. Fortunately, they failed to remove one thing that came in very handy when the R-3350 was faced with serious second-order vibration. Daub had designed the crankcase to be as small as possible. When the design was complete, he lengthened it by two inches. Many saw no need for the additional space, but Daub said, "Don't worry, we will fill it up." Sure enough, when early R-3350s started grinding up reduction gears and shedding accessories, second-order counterbalances, which corrected destructive second-order vibration at the root of this problem, went in this space.[4] Years later after the troublesome debut of the R-3350, engineers faced with the challenge of fixing the engine's many faults went back to Daub's original layouts. In every case, putting back Daub's original features solved the problem.

The Tornado

Daub's execution of the Tornado was nothing short of inspired. According to Daub, the idea of the Tornado was a result of a visit by Gen. H. H. "Hap" Arnold to Wright Aero vice-president Arthur Nutt. Arnold told Nutt he "wanted a 2,000 hp engine no wider than a pilot's ass." In his spare time at home, Daub had worked up the crankshaft and connecting rod geometry for a small-diameter seven-cylinder radial for use in automotive applications. Using this basis along with proven engineering principles and many features from the R-2600, Daub began to flesh out the details of the small-diameter Tornado. A Wright Aero engineer named Willard B. Goodman was involved in initial engine design studies.[5] Both he and Daub were granted patents that detail the mechanical arrangement of the Tornado. When Goodman's patent was issued on January 9, 1945, he was living in Los Angeles, California, and was no longer part of the Wright Aero organization. Goodman must have been instrumental in the creation of the concept, yet the role he played has been obliterated by time.

Two developments proceeded in parallel. A 14-cylinder section of the larger Tornado was fitted with an R-2600 nose case and accessory housing. This prototype engine was used to prove the Tornado concept by running a 50-hour development test and 150-hour endurance test. Meanwhile, the larger 42-cylinder Tornado was built. As with the T-14, initial testing of the T-42 proved the merit of the unusual arrangement of the crankshaft, propeller drive, and accessory drive.

The War Years

Just about the time developmental testing of the T-42 began in earnest, fate intervened with the Japanese bombing of Pearl Harbor and consequent U.S. entry into World War II. Daub had known a number of people in Germany that were associated with engines and racing, and maintained contact with these friends before and after the War. In an attempt to awaken the pacifists into action, he had often predicted that Germany would rapidly and ruthlessly overrun Europe. He had also been instrumental in getting an individual fired from Wright Aero who later showed up in the plant again as an inspector for the Navy. All these otherwise unrelated events conspired against Daub when panic and paranoia swept the nation after the Declaration of War. Daub was brought up on charges of un-American activities and was prevented by the Federal Bureau of Investigation from entering the Wright Aero plant for the remainder of the War. He was never prosecuted and his case never went to court, but the U.S. was deprived of one of its most gifted designers for the duration. Daub had made so many enemies at Wright Aero that only one person stood with him during his troubles—Bill Lundquist.

With access to Wright Aero prohibited, Daub was out of work for much of the War. During that time, he painted a number of pictures, many of which are still in the collection of Ruth Daub, his daughter-in-law.[6] Near the end of the war, he designed a two-stroke engine to power a generator set for the Signal Corps. A prototype was constructed by a machine shop in Belleville, New Jersey. It was smooth running and well balanced because of its careful layout and its use of pumping cylinders for both pumps and balance.

Postwar Years

When the war ended, Daub returned to Wright Aero. Though he continued to be creative and productive, he was never again at the center of the action. One thing Daub did accomplish during his second tour of Wright Aero was to pass on some of his design knowledge. He singled out four young engineers whom he thought held promise and tried to teach them the ropes. These were Stan Baker, Cranston Folley, John Minnich, and Earl Norwood. He taught them the elements of good design and how to recognize its simplicity and elegance. He also showed them examples of poor design.

Daub had little use for magnesium; he commented that it was "like working with cheese." He preferred aluminum. Daub also made an interesting point about the R-4090, an experimental engine with eleven R-3350 cylinders per row: "The angle between the maximum connecting rod excursion and the cylinder centerline is sacred. If it gets too large, there will be trouble. Eleven cylinders per row means long rods and narrow angles. That is one reason why it is so smooth."

Messrs. Minnich and Norwood still recall some of Daub's gems:

> "Don't forget the fundamentals."

> "Easy to design complicated, hard to design simple."

> "We can take lines out of the parts list if you go this way. It will eliminate all the logistics of getting the parts, storing the parts, keeping the specifications up to date, and so forth. Don't put them in if you don't need them."

> "Don't make it holier than the Pope."

> "Every line in a drawing is money."

> "Every specification listed on a drawing is money."

> "Never go with your first idea."

> "It takes more air to cool a liquid-cooled engine than an air-cooled engine."

John Minnich remembers, "Rudy Daub was one of the greatest mechanical designers that I ever knew, and I knew them at Wright Aero, at Allison and at Rolls-Royce. Daub as a mechanical designer, was unequalled."

After the war, the aircraft industry struggled to convert its capacity and capability to commercial endeavors. This was especially true of the engine manufacturers. There was little new design work, especially in the piston-engine arena. Designers who still had jobs found themselves converting wartime engine designs to commercial use. This process consisted mostly of designing product improvements to repair known problems. Daub hated repair work. According to John Minnich, "He wanted to go back and start at the drawing center line* and build it right from the beginning. Of course, you couldn't do this, but in looking at others' designs, Daub often saw that the root of the problem was fundamental and started somewhere near the centerline. While he was good at doing repair work, he did not like it at all."

The rivalry with Chilton continued after the War. Once Daub did a layout of an engine substituting his own design for everything in the engine for which Chilton held a patent. Daub presented his design to the Wright Aero management and pointed out the cost savings. They flatly refused to replace any of Chilton's patented features. Daub later heard a rumor that Chilton was paying kickbacks to Wright Aero management.

During the latter part of Daub's career, he became absorbed with the problems of Diesel engines and piston design. He believed two factors limited the life of Diesel engines – top compression ring groove wear and the cylinder head gasket. He designed cylinders without head gaskets and developed techniques for casting steel inserts into the top ring grooves to improve the performance and durability of the piston and ring. He took his ideas to one of the foremost piston manufacturers, Zollner Pistons. Daub arrived at the gate without an appointment. Within ten minutes he had talked his way into the

*In the days of drawing boards, the center line was the first line to be drawn on any drawing.

office of the company founder, Fred Zollner. By the end of the morning, he had sold the concept. Daub was still active in piston design issues when he died on February 7, 1975.

Bill Lundquist

Wilton G. "Bill" Lundquist was the engineer in charge of the Tornado program in its later days, after the departure of Goodman. It is difficult to quantify Lundquist's direct contributions to the Tornado. As we shall see later, he was heavily involved at the program level. Some of his patents, especially those for dual-rotation and two-speed reduction gear arrangements, would have seen service on the Tornado.

Like his direct contributions to the Tornado, the personality of the man is equally obscure. Lundquist was well-liked, and many called him "Lundy." Nicknames, at least those called to their face, were uncommon for managers in the stuffy corporate culture of Curtiss-Wright. Guy Vaughan, president of Curtiss-Wright during the Tornado program, called him "open-minded, aggressive, and intelligent." Dr. Arthur Nutt, vice-president of engineering at Wright Aero, called him "a very capable and brilliant engineer." While nearly everyone who worked for Wright Aero remembers him, few remember much *about* him.

With patents for such things as selective speed propeller drive, automatic manifold pressure control, crankshaft vibration dampers, innovative engine arrangements, and torquemeters, it is obvious that Lundquist was involved in hands-on engineering activities as well as engineering management. For decades following, other inventors referenced his November 7, 1944 patent for an improved engine cooling system. His greatest contribution to Wright Aero, and to the art of engine design, may have been in his skill at putting capable people in key positions and giving them the freedom to do what they did best. One such example was Rudy Daub.

John Minnich

John Minnich was a test engineer with the Tornado project when he first came to Wright Aero. John was born in Zanesville, Ohio (about 50 miles east of Cleveland) on April 23, 1914. He attended Zanesville High School. Like many kids at that time, he was interested in aviation. After graduating from Ohio State University in 1941, Minnich was recruited by Wright Aeronautical. Along with 60 to 70 other young engineers, he began an apprenticeship in the test department where he remained for about one year. The pay was $0.65 per hour.

John worked for fifteen weeks on the dynamometer stand where the Tornado was tested. It took almost a full shift to set it up on the test stand. Forty-two thermocouple leads to the spark plugs, and numerous pressure and temperature sense lines for oil and coolant had to be hooked up. Apprentice test engineers installed the engine in the test cell during one shift, ran it the next, and removed it on the third. When there was no engine to test, they got the duty of mopping the oil from the floor and walls of the test cell. The first runs were done on a dynamometer. Later runs were done outside on the flight propeller test stand.

John's recollection of running the Tornado is still vivid today.

> When you pressed the throttle forward, the rpm was right there. I never saw anything that accelerated or decelerated as fast as it did. When it was running in the test cell, short exhaust stacks were fitted instead of the usual exhaust system. In the darkened test cell, it was a spectacular sight to see 42 plumes. By changing the fuel/air ratio, the flames could be varied from short gray to a long, bright blue. There were problems with unequal air distribution. Cylinders two and five [of each bank] got twice as much air as the others.

Minnich also recalls how Carl Stuart, another test engineer, accidentally found a weak point in the engine: "He was the test engineer on a Tornado run late one night. He nodded off and his head fell to the kill switch. As soon as the engine sound disappeared, Stuart awoke and pulled the kill switch back out, restarting the engine. This immediately sheared the supercharger impeller shaft, resulting in the redesign of the impeller drive. Happy accidents such as this were commonplace during engine development."

Although he worked on pistons, rings, and valves at Wright Aero, Minnich is most proud of work on piston rings. He authored a paper entitled "Development of Chrome Plated Piston Rings for Aircraft Engines" that was presented to the Society of Automotive Engineers on October 4, 1946. He helped perfect chrome plated wedge-shaped compression rings with tapered faces that combined favorable wear qualities with good oil consumption.

Around 1947, Curtiss-Wright had a huge layoff. In one two-week period, two-thirds of the engineering staff was released. "An optimist was one who brought his lunch to work," recalls Minnich, who managed to survive that particular reduction in force. When it became clear in 1956 that the investment bankers were going to squeeze Curtiss-Wright dry, Minnich moved to Allison where he worked on joint programs with Rolls-Royce. He was twice assigned duty as an Allison engineering representative in England. The partnership was trying to get engines into the Boeing 727, and later, the General Dynamics F-111. Neither program was successful. In 1968, Minnich worked on a lift engine for a VTOL application and succeeded in producing an engine with a ten-to-one thrust to weight ratio. This was an amazing figure for the time.

Design Influences and the Tornado Patents

Good design steals from others, improving on what has passed the test of time. Such was the case of the Tornado. A review of the Tornado patent literature elucidates the evolution of ideas and illuminates accomplishments of some designers and engineers who worked on it.

W. B. Goodman is credited with the idea of piloting the aft end of one crankshaft into the front end of the one immediately behind it. In this manner, slight relative motion between the two shafts was permitted. Another advantage was that a single main bearing could be shared between two separate crankshafts. This made the engine shorter, lighter, and stiffer.

Goodman is also credited with the overall engine arrangement; namely multiple rows of in-line cylinders arranged radially, with a gear-driven overhead camshaft in each row. Finally, it was Goodman who patented the three tandem crankshafts geared together with multiple high-speed layshafts. It is possible this idea was borrowed from a German designer, Fritz Gosslau. Gosslau applied for a U.S. patent on October 12, 1938, and was granted U.S. Patent No. 2,297,314 on September 29, 1942. Gosslau had also applied for a German patent on October 14, 1937. Gosslau's design never showed the connection of more than two crankshafts, but the idea of a large crankshaft gear driving multiple layshafts was his. The Tornado concept accommodated one, two, three, four, or five crankshafts. Daub, with all his German connections, may have known of Gosslau's work.

Daub and Goodman must have collaborated extensively on the Tornado. Their two patents, 2,355,277 and 2,366,852 were both filed on December 27, 1941, and bear application serial numbers 424,560 and 424,563, respectively. Both their patent drawings were obviously drawn by the same artist (Daub?) and were processed by the same patent attorney.

Daub's claims were much more expansive than Goodman's were. In his U.S. Patent No. 2,355,277, Daub claims as original his method of constructing the crankcase, and his design for the cooling system and coolant flow. Daub further claims the application of this cooling system both to the general class of engines with the arrangement of the Tornado (multiple in-line blocks of cylinders arranged radially around a common crankcase), and to engines with the specific construction of the Tornado (multiple cylinder blocks bolted to a polygonal tubular crankcase.)

In a later patent, 2,511,248, issued June 13, 1950, Daub claims as original his approach to the lubrication of valve gear through a hollow camshaft that includes a means of controlling the amount of oil delivered to the valve gear. This is accomplished by reducing the oil pressure within the camshaft to a value below that supplied by the engine pressure pump. He further claims the design of the oil pressure reducing apparatus and the design of the reducer valve in general.

While it is clear that many innovations were Daub's, other influences are just as clear. The Tornado torquemeter is an elegant repackaging of an earlier design by Roland Chilton (U.S. Patent No. 2,289,285). The intake manifold is taken from Glenn D. Angle (U.S. Patent No. 1,622,272). Other contributors include Loomis, White, Brown, Vincent, Noble, Schenk, Farina, Woolson, Codrington, and Davis.[7]

While we are able to trace some influences via patents and historical accounts, there are doubtless others that took place over cups of coffee and lunch pails, ones that have forever faded with the memories of the participants.

---

1 Unless otherwise noted, all biographical material about Rudolph Daub comes from telephone interviews of John Minnich by the author during December 2000 and January 2001.

2 Ruth Daub, telephone interview by the author, December 17, 2000.

3 Ibid.

4 Earl Norwood, telephone interview by the author, December 17, 2000.

5 Goodman, *Flat and Radial Engine Design Studies.* See also Goodman, *Tornado Engine Model Variations.*

6 Ruth Daub, telephone interview by the author, December 17, 2000.

7 Daub, U.S. Patent No. 2,511,248.

3 The Tornado 14

The Tornado 14 was a two-row prototype module of the larger six-row Tornado 42. (Courtesy of the New Jersey Aviation Hall of Fame and Museum)

Based on the success of the R-2160 proposal, the Army gave Wright Aero a contract for an experimental fourteen-cylinder section of the engine on June 24, 1939.[1] Two of these engines, known as the Tornado 14 (T-14), were built. [2] They were Wright Aero engine serial numbers 11760 and 11761. It is probable that serial numbers 11762 and 11763 were also assigned to T-14s, as enough T-14 parts were built for four engines. There is no evidence, however, that the last two T-14s were ever built.

The T-14 was never intended to be a complete engine or to fly. Instead, it was the basic power module of proposed larger Tornado engines, and was to prove the feasibility of features unique to the Tornado. Study of the T-14 is important for several reasons. First, since it completed a 50-hour development test and a 150-hour endurance test, two surviving reports[3] provide a wealth of information about the engine. Many of these details are applicable to the Tornado 42. Second, the T-14 shows remarkable design savvy in the way that it made use of mature components from already-developed Wright Aero engines. Finally, some of the problems identified with the T-14 were to plague the program for its duration. It is instructive to analyze these problems and ponder why they were not given more attention early in the program.

This chapter presents an overview of the construction and materials used in the T-14, photographs of essential components, and brief accounts of the formal tests. Many more details are available in Appendix B—"Tornado 14 Detailed Description" and Appendix C—"Tornado 14 Tests."

Front View

Rear View

Top View

Bottom View

By combining new concepts with proven components, the Tornado 14 tested the basic engine power section. (Courtesy of the New Jersey Aviation Hall of Fame and Museum)

Description of the Tornado 14

The T-14 was a fourteen-cylinder, two-row, liquid-cooled static radial engine. It operated on the conventional four-stroke cycle and used aviation gasoline for fuel. With a bore of 4.250 inches and a stroke of 3.625 inches, it displaced 720 cubic inches and produced 600 horsepower.

Since the T-14 was built primarily for development testing of the power section, the nose case, supercharger section, and rear accessory section of the T-14 were adapted with as little change as feasible from the Wright C14A (R-2600) engine. This reduced both the risk and level of effort by applying existing, proven technology where practicable. The power section was almost completely new.

The Power Section

Most multi-row air-cooled engines stagger the rows so that the rear cylinders are between the front ones. With the T-14, seven banks with two in-line cylinders each were arranged radially around a central crankcase. The crankcase housed the crankshaft unit, layshafts and connecting rods. At the front of the crankcase was a front accessory case that housed camshaft and propeller drive train components. At the rear of the crankcase was an adapter section that housed supercharger and accessory drive train components.

Crankcase

The crankcase consisted of a front, a center, and a rear crankcase assembly bolted together by 28 bolts, two of which passed fore and aft through the space between each cylinder bank. The front crankcase assembly attached to the front accessory case. The rear crankcase assembly attached to the adapter section. All finish-machining operations on the crankcase were done with the crankcase assemblies bolted together as a unit to ensure good alignment. Provisions existed within each crankcase assembly for a crankshaft main bearing bushing.

Seven layshafts occupied the space between adjacent cylinder banks and transferred crankshaft power to the propeller and engine accessories. Bushings were provided in the front accessory case, front crankcase assembly, rear crankcase assembly, and adapter section for each layshaft. All layshaft bushings were line bored to within 0.001 inch of one another to ensure proper layshaft journal wearing qualities.

The cylinder banks were cast of bronze with shrunk-in steel cylinder liners, steel valve seat inserts, and bronze valve guides. Bronze was probably used in this experimental engine because it was easier to cast in large sizes than aluminum. At least one full sized Tornado T-42 was produced with bronze cylinder banks. This was the R-2160-1 designation. These were later retrofitted with aluminum

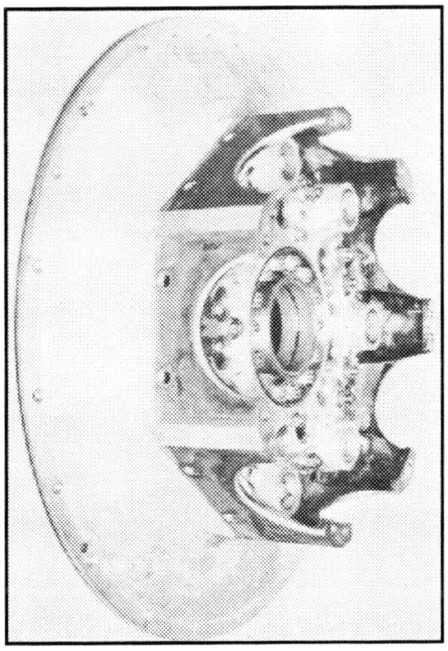

T-14 Front Crankcase Assembly (New Jersey Aviation Hall of Fame and Museum)

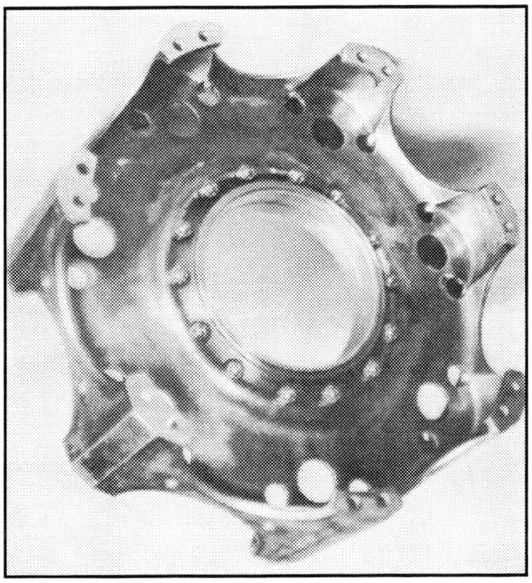

T-14 Center Crankcase Assembly (New Jersey Aviation Hall of Fame and Museum)

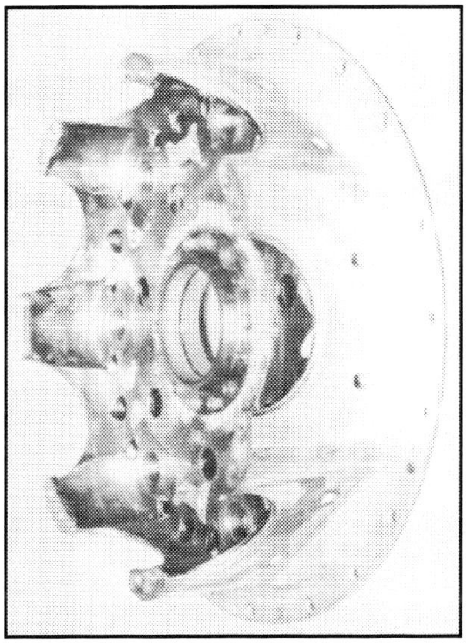

T-14 Rear Crankcase Assembly (New Jersey Aviation Hall of Fame and Museum)

banks, and became the R-2160-3 designation. The cylinder liners were machined from alloy steel and then nitrided to improve wear qualities. The liners were then installed into the cylinder bank and the inside diameters honed to their final size. Coolant jackets surrounded the cylinder liners. Both sides of the cylinder bank were covered by side cover plates screwed to the bank casting. Coolant entered at the rear top of each bank and exited at the front top.

Each cylinder was fitted with one exhaust and one intake valve located in the hemispherical dome on opposite sides of the combustion chamber. The sodium-cooled exhaust valves were made of corrosion and heat resistant steel with stellite faces. The intake valves were of heat resistant steel and were formed with tulip-shaped heads and hollow stems. Each valve was fitted with three valve springs, locks, and spring washers of conventional construction. The valves were operated through rocker arms running on an overhead camshaft with an individual lobe for each valve. Valve clearances were adjusted via an adjustment screw on the valve end of each rocker arm.

The camshafts were driven by gears in the front accessory case and ran in half-bushings captured between the cylinder head and camshaft bearing caps. Camshaft bushings were line bored to ensure proper alignment. Cylinder covers were of cast aluminum and enclosed the entire top end of each cylinder bank.

The fuel-air mixture was supplied to the cylinders through cast aluminum manifolds attached to the intake ports by screws. These manifolds were attached to the supercharger front housing outlets by rubber tubing and clamps.

Crankshaft

The crankshaft unit was a two-throw, three-piece assembly that followed standard Wright Aero practice of using built-up crankshafts and one-piece master rods. It rotated in a counterclockwise direction when viewed from the rear. The three crankshaft sections (front, center, and rear) were machined from alloy steel forgings, and each was drilled to reduce weight and provide oil passages. Standard Wright Aero clamp joints were used to attach the front and rear crankshaft sections to the center crankshaft section.

The center crankshaft section consisted of two crankpins 154 2/7° apart separated by a journal for the center main bearing. Bolted to either side of the center main bearing journal were master rod spacer plates that served the dual role of controlling the master rod position on the crankpin and the crankshaft unit end play. Within the drillings of the center main bearing journal and each crankpin were aluminum plugs that formed oil passages within the crankshaft center section. These plugs could be extracted to facilitate the removal of sludge separated from the oil by centrifugal force.

The 154 2/7° angle between the crankpin throws is noteworthy. Most air-cooled, two-row radial engines have cylinders in the rear row directly between those in the front row. This construction facilitates cooling and provides even firing intervals for all cylinders when the front and rear crankshaft throws are spaced 180° apart. The Tornado differed from this construction in that the rear cylinders were directly behind the front ones. Since it was liquid-cooled, delivery of cooling air was not a problem, but there remained the issue of how to achieve even firing intervals. This was accomplished by spacing the crankshaft throws at 154 2/7°. Since the Tornado had seven cylinders per row, a 154 2/7° crankshaft was mechanically equivalent to staggering the cylinder rows by an angle of 25 5/7° (one-half of the angle between each of seven cylinders).

Front and rear crankshaft sections were attached to the center crankshaft section via clamp joints. Each front and rear crankshaft section consisted of a crankcheek, a main bearing journal, a counterweight and a dynamic damper. The crankcheeks were bored to fit tightly over the crankpins and split through the top, forming a clamp. A large clamp bolt tightened to a predetermined stretch provided enormous force to close the clamp around the crankpin. The crankshaft sections were aligned during

T-14 Crankshaft (Wright Field Technical Library Collection, Acc. XXXX-0428, National Air and Space Museum, Smithsonian Institution, SI Neg. No. 2001-834

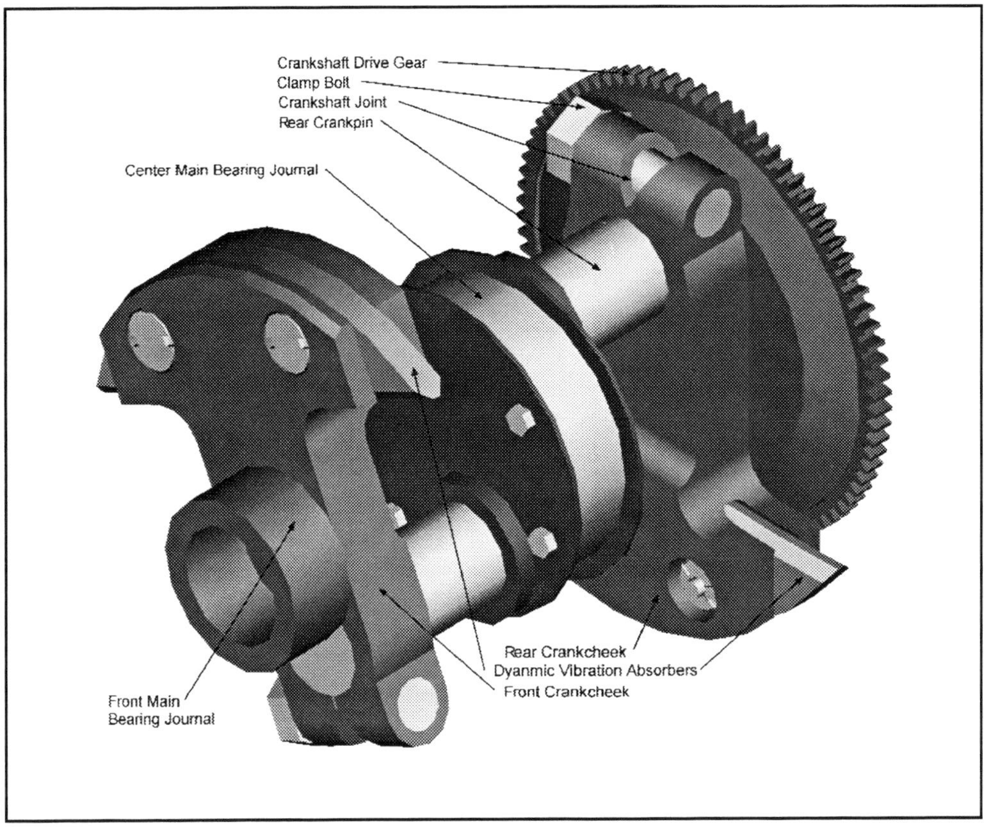

Left Front View of the T-14 Crankshaft

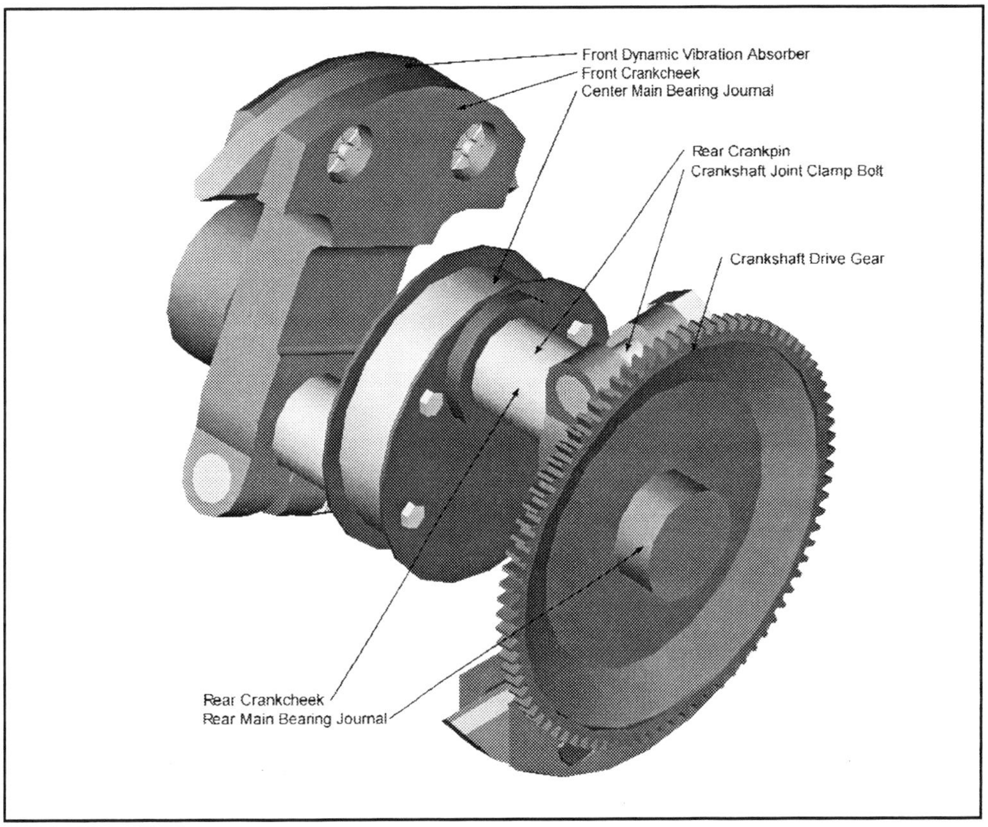

Left Rear View of the T-14 Crankshaft

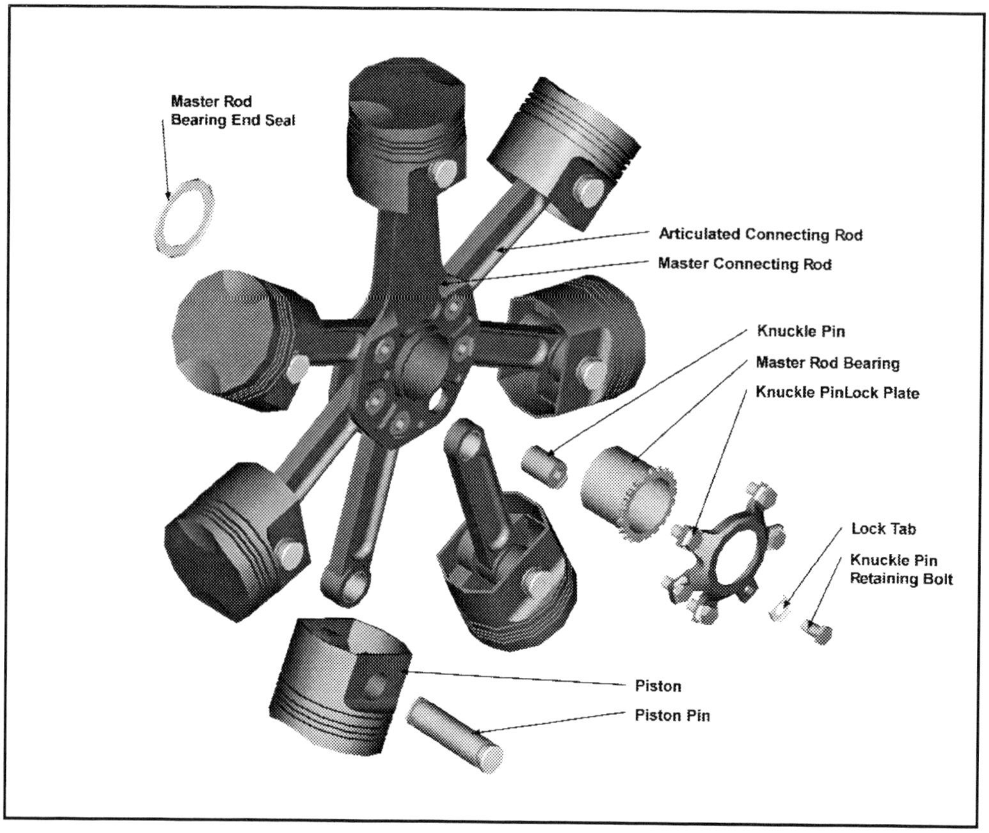

Exploded View of the T-14 Connecting Rods and Pistons

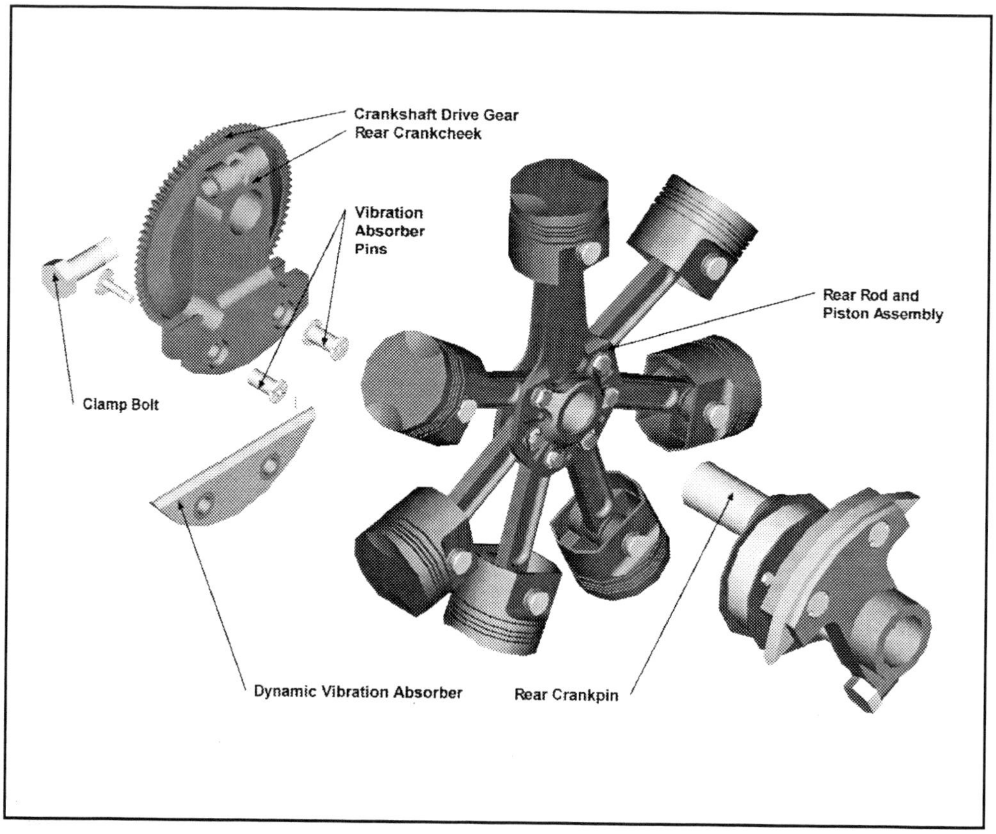

Exploded View of the T-14 Crankshaft, Connecting Rods, and Pistons

assembly and checked for alignment at each disassembly.

As was the usual practice, counterweights were sized to offset all of the rotating mass and one-half of the reciprocating mass of the crankshaft, connecting rod assembly, and pistons. The dynamic dampers were bronze weights supported on the crankcheeks by floating pins and tuned to dampen second order torsional vibration. A crankshaft drive gear bolted to the rear crankshaft section drove the layshaft pinions and acted as the power take-off for the crankshaft.

Connecting Rods

The connecting rods followed standard Wright Aero practice, except they were shorter than usual to reduce the engine diameter. The master rods were machined from alloy steel forgings. The crankpin end was channel sectioned for the attachment of articulated rods, and the shank section extended up from the channel section as an "H" section to the piston pin eye.

The articulated rods were machined from "I" section alloy steel forgings. Knuckle pins pressed into holes in the master rod channel section attached the articulated rods.

Master rod bearings were of steel-backed copper lead pressed into the rod before final boring and lead plating. Piston pin and knuckle pin bushings were the split bronze type pressed into place and honed after installation. End seals on the master rod bearing ensured a pressure supply of oil to the knuckle pins. Knuckle pins were lubricated by oil passing through the knuckle pin lock plate to the knuckle pins.

Pistons

The pistons were made from aluminum alloy forgings. They were designed with flat heads except for valve reliefs. The ring pack was made up of three wedge-type compression rings and one flat oil control ring, all located above the piston pin. Oil vents to the inside of the piston were provided between the bottom and third ring grooves. Rings were cast iron with tapered outside diameter faces to scrape oil down the cylinder liners. The pistons were joined to the connecting rods by full floating highly finished steel piston pins. The pins were held in place by dural plugs which pilot in the pin and ride against the cylinder walls.

Layshafts and Associated Pinions

A large gear at the rear of the crankshaft drove the layshaft driven pinions at three times crankshaft speed. Small pinions at the rear of each layshaft driven pinion supplied power to the accessory gears. Internal splines within the front of each

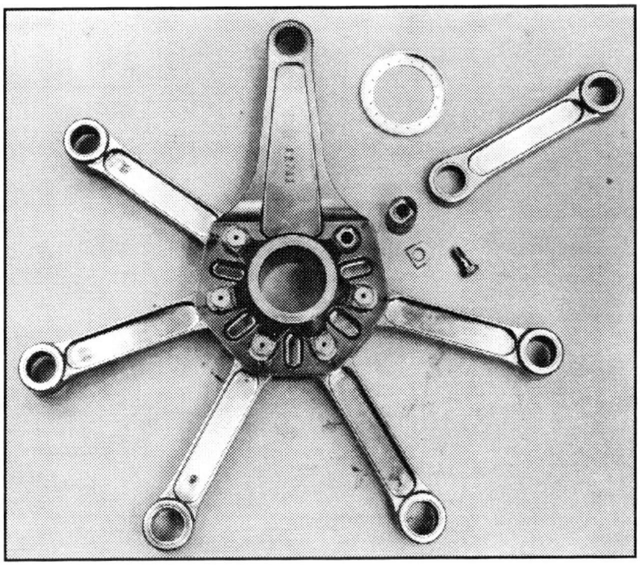

T-14 Master and Articulated Connecting Rods (Wright Field Technical Library Collection, Acc. XXXX-0428, National Air and Space Museum, Smithsonian Institution, SI Neg. No. 2001-836)

Tornado Piston (Courtesy of the New Jersey Aviation Hall of Fame and Museum)

layshaft driven pinion mated to splines on the rear of each layshaft. The hollow layshafts carried power forward to the front accessory drives and propeller reduction gearing. Each layshaft had two integral pinions at its front. The most forward pinion supplied power to the camshaft drive train. The larger pinion meshed with an intermediate power gear that was part of the propeller reduction gear.

Front Accessory Case

Two idler gears and one compound idler per camshaft were housed within the front accessory case. The front scavenge pump was driven from a camshaft driving gear through a spur gear idler. All cam and pump drive gears were involute spur gears formed from steel alloy forgings, hardened, and ground.

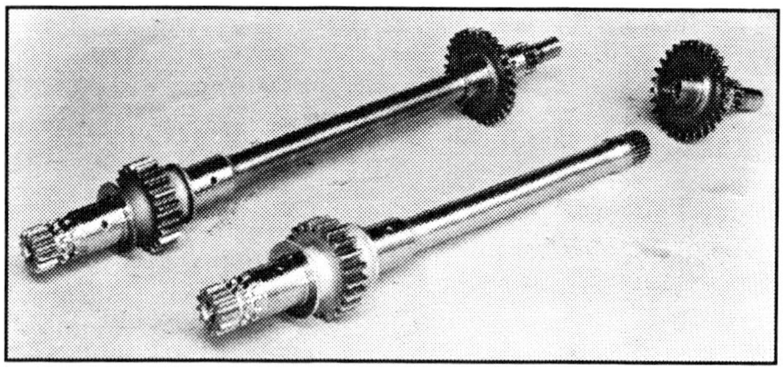

Layshafts (Wright Field Technical Library Collection, Acc. XXXX-0428, National Air and Space Museum, Smithsonian Institution, SI Neg. No. 2001-835)

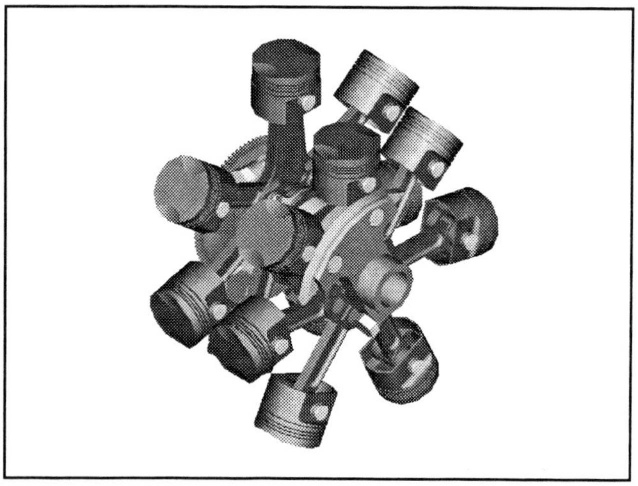

Complete T-14 Crankshaft Assembly

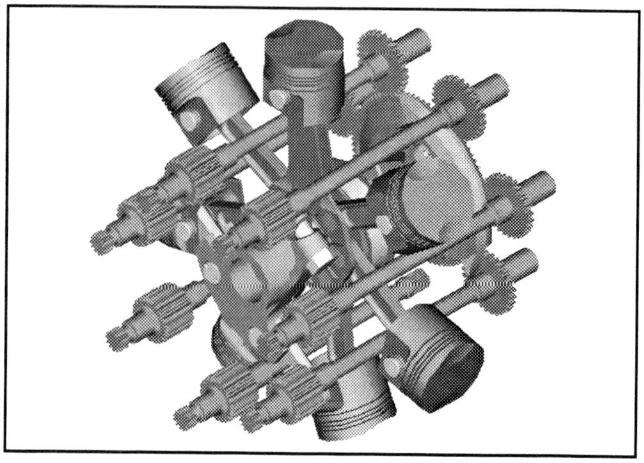

The T-14 Power Section, with Rear Cylinder Row Omitted, Showing the Relationship Between the Crankshaft Assembly and the Seven Layshafts

The Nose Section

While use of the C14A nose section greatly reduced the overall level of effort, there was still some clever design involved in making the T-14 power section work with it. In the C14A, a long crankshaft extension forms a pilot for the planetary reduction gear and propeller shaft. Splined to the crankshaft extension is a ring gear through which power is supplied to the reduction gear. The T-14 crankshaft did not deliver power directly to the reduction gear. Instead, the layshafts were machined with a 21-tooth reduction gear drive pinion located at the front. These pinions drove a 91-tooth intermediate power shaft gear. Splined to the intermediate power shaft was the usual C14A ring

The C-14A crankshaft had a long front extension that served as a pilot for the propeller shaft and reduction gear. (Courtesy of the New Jersey Aviation Hall of Fame and Museum, annotated by the author)

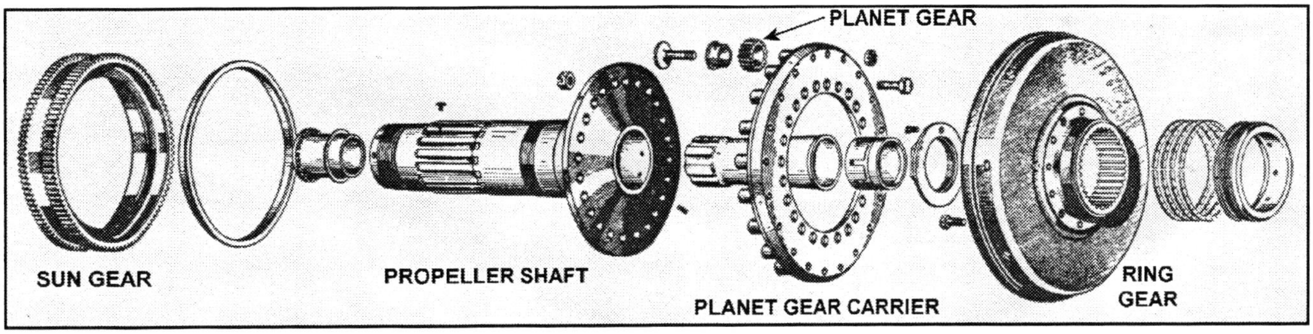

The large ring gear of the stock C14A reduction gear was splined to the back of the crankshaft extension. The propeller shaft was driven by the planet gear carrier and was piloted on the front of the crankshaft extension. The sun gear was fixed to the nose case. Twenty planet gears meshed with both the sun gear and ring gear. (Courtesy of the New Jersey Aviation Hall of Fame and Museum, annotated by the author)

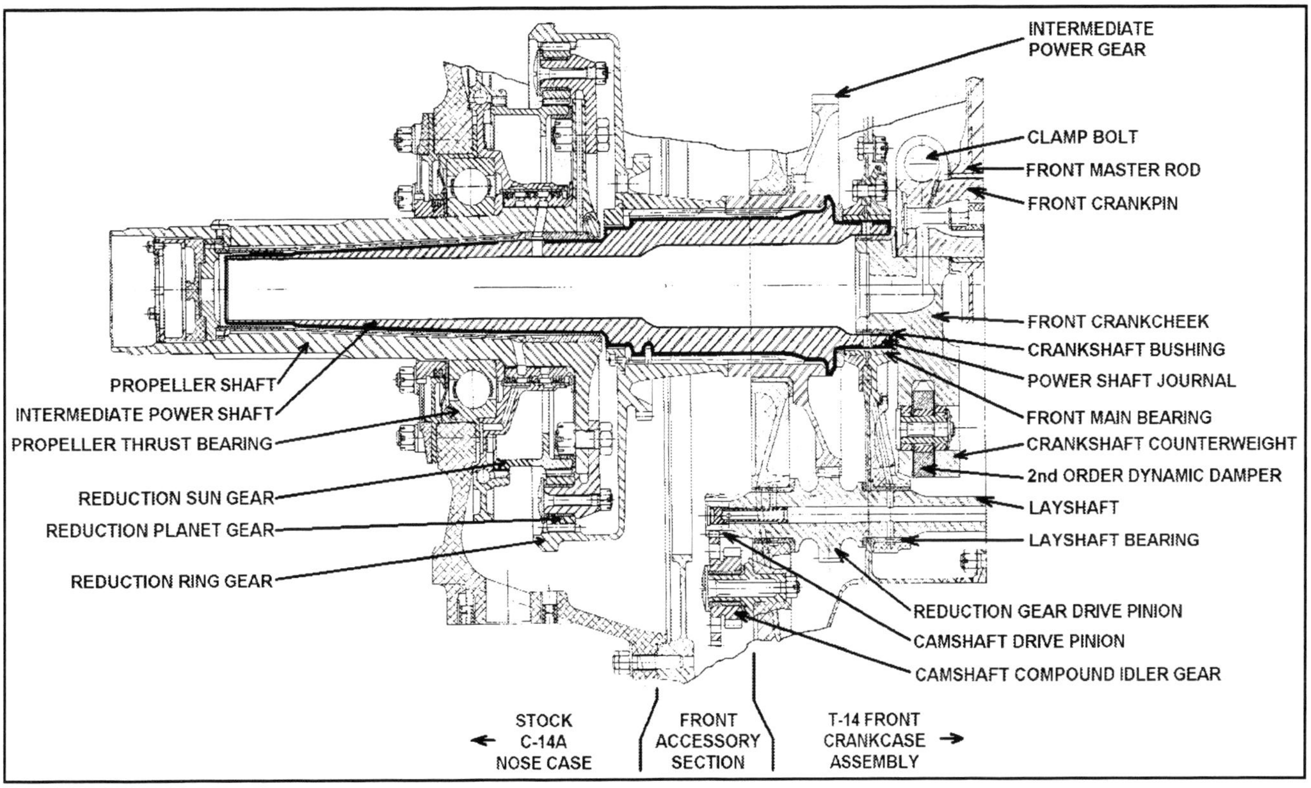

Partial Cross-section of the T-14. An intermediate power shaft (heavy outline) was driven by the intermediate power gear. On the same set of splines was a stock C14A ring gear. Note the rear end of the intermediate power shaft runs in the front main bearing, while the crankshaft front main bearing journal runs in a sleeve inside the intermediate power shaft. Bearing provisions for the front of the intermediate power shaft duplicate those of the C14A. (From Wright Field Technical Library Collection, Acc. XXXX-0428, National Air and Space Museum, Smithsonian Institution, file D45.41 Wright/584 (WAC Report 532, drawing #413940), redrawn and annotated by the author)

gear. This additional gear stage not only allowed the layshafts to drive the reduction gear, but also provided additional speed reduction to account for the higher crankshaft speed of the T-14.

A bearing journal at the rear end of the intermediate power shaft ran in a bushing pressed into the front crankcase assembly. In the rear end of the intermediate power shaft, a bushing in a counterbore formed the front crankshaft main bearing. The front end of the intermediate power shaft ran in bushings inside the propeller shaft. The propeller thrust bearing controlled fore-and-aft movement of both the propeller shaft and intermediate power shaft. The thrust of the intermediate propeller shaft was taken by a plain thrust bearing in the front accessory case.

The nose section was modified to deliver the entire oil supply to the propeller shaft. Since the T-14 was not a flight article, there was no need for a separate high-pressure oil supply to the propeller.

The C14A Accessory Drive Gear Train (Courtesy of the New Jersey Aviation Hall of Fame and Museum)

The Adapter Section

The adapter section was a cast aluminum case adapting the crankcase to the C14A supercharger front housing and rear accessory section. It supported the intermediate accessory drive gear and rear accessory section drive pinions and provided a coolant inlet header to the cylinder banks.

Power from the crankshaft was delivered to the rear accessory section through 12-toothed pinions that were integral with the layshaft driven pinions. These pinions drove seven 32-tooth idler gears, which drove a central 72-tooth intermediate accessory drive gear. The accessory drive gear piloted in the crankshaft rear journal and was splined to the accessory drive shaft. The T-14 accessory drive was designed to run the accessory drive shaft at one-half crankshaft speed, which was one-half the normal speed of the C14A accessory drive. This was done to prevent supercharger overspeed. Since the T-14 crankshaft rotated at nearly twice the speed of the C14A, it was necessary to halve the supercharger drive ratio. A consequence of this was that the magneto drive train had to be changed to preserve the 0.875:1 magneto drive to crankshaft ratio.

Supercharger Front and Rear Housings

The supercharger front and rear housing assemblies were standard Wright C14-A two-speed sections except for minor changes.

Mounting

Four mounting pads were attached to the supercharger front housing. The mounting pads were bolted to an engine mount.

Breathing

Crankcase ventilation was through the supercharger front housing from the rear sump. Relief valves in the upper scavenge passages eliminated excess crankcase pressure in the event the crankcase became loaded up with oil.

Cooling System

A coolant mixture of ethylene glycol and water entered the engine through a centrifugal coolant pump mounted on the generator pad. Coolant was pumped through two external lines to a rear coolant annulus in the adapter section. This annulus acted as an inlet header and delivered the coolant to the rear top of each cylinder bank through short connecting tubes. After passing through the banks, coolant exited from the front top of each bank into a front coolant annulus in the front accessory case. Coolant left the engine through a fitting in the front coolant annulus. The test operator was responsible for control of coolant temperature and flow rate.

Lubrication System

Like most radial engines, the Tornado 14 relied on oil under pressure to lubricate most of the bearings and journals, and splash oil to lubricate the remainder of the moving surfaces. Once the oil had served its lubrication function, it was collected by scavenge pumps and returned to the oil supply tank.

Pressure Circuit, Rear Branch

Oil from an external supply tank entered the engine oil pressure pump and was pumped through an external filter to layshaft oil pressure passages via two external lines connected to fittings on the adapter section. One line fed the #3 layshaft and the other the #4 layshaft. From this point, oil flowed forward through the hollow layshafts, and aft through drillings in the adapter section. From the adapter section, rear accessory section lubricating oil flowed through the intermediate accessory drive shaft and accessory drive shaft to the rear accessory section.

An orifice in the rear accessory drive shaft reduced the pressure and restricted the flow to the rear accessory section bushings. Contrary to usual practice, engine lubricating oil was not used to control the supercharger clutches. Instead, auxiliary oil pressure was supplied to the supercharger clutches through a rear cover pressure fitting. The oil pump was blanked off from the rear accessory section and delivered oil externally to the adapter section at two places.

Pressure Circuit, Front Branch

Oil to the engine sections forward of the rear layshaft support was delivered to the layshafts through the rear accessory section drive pinions. From the layshafts, oil traveled through the layshaft bearings and supports to the rear crankshaft bearing, the front crankshaft bearing, the front accessory case, and the nose case oil passages.

Oil under pressure to the front and rear main bearings was distributed through drillings in the crankcheeks and crankpins to the center main bearing, the master rod bearings, and the knuckle pins. No jets were provided in the shaft to increase the crankcase oil supply because the leakage from the crankpin bearings provided sufficient lubrication for the cylinder walls, pistons, rings, piston pins, dynamic dampers, and gears. Oil was also supplied to the front crankcase and reduction gear assembly from the crankshaft through the intermediate power shaft.

The front accessory case was drilled to supply oil from the front layshaft supports to the cam drive gear train journals. From there it flowed through the camshafts to the camshaft supports and rocker assemblies. Oil was delivered to the intermediate propeller shaft from the layshafts through the front crankcase assembly layshaft support.

Scavenge Circuit

The crankcase was scavenged by gravity drain to the lower part of the engine. The front accessory case, the nose case, and the front of the crankcase drained directly to a gear type front scavenge pump mounted on the front accessory case. The rear of the crankcase, the adapter section, and the front and rear supercharger housings drained into a sump and was removed from the engine via the rear scavenge pump. Each camshaft cavity was scavenged by individual gear-type scavenge pumps driven off the rear of the camshafts. Scavenge oil under pressure was then pumped to the front accessory case scavenge annulus. From there, it flowed to the front scavenge pump outlet chamber and from the engine. All scavenge oil was collected and routed through a heat exchanger, filter, and finally back to the oil supply tank.

Ignition System

The engine was equipped with two Scintilla SF-14L magnetos running at 0.875 times crankshaft speed. These magnetos were the same type used on the C14A engine. The ignition harness was completely shielded. Dual ignition was used, with each magneto firing one of the two spark plugs in each cylinder independently. Timing for each magneto was fixed at 30° BTDC.

The 50-hour Development Test

The 50-hour Development Test was a contractual milestone intended to demonstrate progress and identify problems. This testing was performed on Engine No. 11761 from July 23, 1940 through September 23, 1940, in partial fulfillment of Army Air Corps Contract W535 AC-12632. The engine had been built up in the Experimental Assembly Department of Wright Aeronautical Corporation. Prior to the endurance test, the engine had accumulated 18:50 hours. The actual test consisted of five break-in hours at 500 hp/3,800 rpm, ten five-minute take-off power periods of 600 hp/4,150 rpm, 28:45 hours at a cruise power setting of 450 hp/3,670 rpm, and 6:50 hours of miscellaneous running. A full copy of the 50-Hour Development Test Report appears in Appendix C—"Tornado 14 Tests."

Test Rig Description

All of the testing was done on Wright Aero test stand X-10 in Paterson, New Jersey. Power was absorbed and measured by means of a dynamometer and a water brake. This was a typical test setup for the time. By using a water brake of known capacity to absorb most of the engine power, a smaller (and less expensive) dynamometer could be utilized to absorb the remaining power. In addition to its role of absorbing power, the dynamometer could also be used as a motor to rotate the engine for starting or non-firing tests. The water brake was rigidly connected with the dynamometer so that all torque taken by the water brake was transmitted to the dynamometer torque arm and was there measured by a scale. By knowing the torque produced by the engine and the speed of revolution, it was possible to calculate the horsepower being produced.

Coolant from a coolant tank flowed to the inlet of the engine-driven coolant pump. After serving its purpose of cooling the engine, it passed through a heat exchanger and back to a coolant tank. This allowed precise control of the coolant inlet temperature. Coolant flow rate was measured as well. Coolant flow was controlled by a throttle valve in the coolant exit plumbing external to the engine.

Engine oil was pumped from a supply tank to the engine via the engine-driven oil pressure pump. Oil leaving the engine scavenge pumps passed through heat exchangers and filters before returning to the tank. Oil flow was measured at the oil tank suction line.

Induction air was drawn into an air plenum through a fixed orifice by a centrifugal blower and forced through heating coils to the carburetor intake. The quantity of air supplied to the engine through the orifice was measured by a draft gage. Using this test setup, a known quantity of air could be supplied to the engine, and the carburetor inlet temperature and pressure carefully controlled. Air entered the engine through a down draft carburetor, then passed through the supercharger impeller and the diffuser section into the intake manifold. From there it was distributed to seven intake pipes, each of which supplied one cylinder bank.

Fuel passed through a heat exchanger and was heated via steam or cooled via water as required. Fuel flow was measured by timing the consumption of a given weight of fuel.

Development Test Summary

After the first part of the development test, five layshaft front bushings were found seized to the shaft journals. During this part of the testing, much difficulty was encountered in maintaining desired oil pressure at various points throughout the engine. Before the second part of the endurance test, the low oil pressure condition was corrected by plugging some crankshaft oil passages unintentionally left open during assembly and by distributing the oil from the engine pressure pumps directly to the layshaft oil passages. The replacement front layshaft bushings were lead-indium plated.

During the second part of the endurance test, one layshaft front bushing seized to the shaft journal and one layshaft rear bushing and journal were deeply scored. This was thought to be due to a drop in oil pressure because of a low oil supply. Upon completion of the 50-hour test, layshaft bushing clearance was increased from the initial specification of 0.0015" – 0.003" to a new specification of 0.003" – 0.0045". It was believed this additional clearance would further reduce the possibility of bushing failure.

After the first part of the endurance test, the master rod bearings were found to be slightly picked out near the edge. This was blamed on a process problem encountered by the bearing vendor. New bearings with lead-indium plating were installed for the second part of the endurance testing and proved satisfactory. Improvements in oil distribution may have also contributed to the improved condition of the master rod bearings during the second part of endurance testing.

Other difficulties were encountered during the testing, but were minor in nature. The number ten camshaft bearing and journal were deeply scored because of dirty oil. The ball retaining wall of the #3 exhaust valve clearance adjusting screw was chipped because of a flaw in the material of the part. An intermediate power shaft bushing turned and backed out 0.25 inches from its position on the shaft, resulting in a design change to a floating bushing. Small cracks in two intake valves were traced to manufacturing flaws.

The 150-hour Endurance Test

The 150-hour Endurance Test was another contractual milestone. Had the engine been a flight article, successful completion of this test would have usually qualified it for subsequent flight testing. This test was performed on engine 11761 from March 1 to June 14, 1941. Prior to the start of this test, the engine had been rebuilt seven times and had accumulated 157:25 hours running time. The test differed from the usual 150-hour endurance test in that no attempt was made to complete the test in one installation of the engine. Instead, Wright Aero engineers reserved the right to disassemble and inspect the engine anytime they felt the need. All running was done on a dynamometer stand rather than a propeller stand.

The engine was removed from the test stand and partially or completely disassembled and inspected at intervals of from 20 to 50 hours. Parts were replaced where necessary and any later or better designs were incorporated into the engine at these inspections. Due to the number of removals, all minor changes are listed in Appendix C—"Tornado 14 Tests" along with some explanation of the reasoning behind a change.

Endurance Test Summary

Operation of the engine throughout the test was good. It was easily controlled and produced the expected power. While trouble with individual parts was experienced, no major weaknesses were encountered. One articulated rod broke during testing and resulted in considerable engine damage. These rods were known to be weak, and once stronger rods were installed, no further rod difficulties were encountered.

Six master rod bearings were replaced during the test because of cracking and picking out of the bearing material. None had caused any damage to other parts. All were part of a batch with inferior bonding and structure. Rather than wait for bearings of better quality from their bearing vendor, Wright chose to continue testing with the bearings already procured. It was anticipated that the vendor would solve its process problems and ultimately produce bearings of the same quality as that for other Wright engines.

A total of six rear bank hold down studs were broken at the bottom thread of the nut. These studs were redesigned to give a better load distribution by burying the entire thread in the nut. Since larger studs were to be used on the T-42, stud breakage on the T-14 was not thought to be significant to the T-42 program.

Replacement of both intake and exhaust valves was necessary because of heat damage due to excessively lean mixtures during run-in. A number of valves that were rejected during magnaflux inspection and removed were later found satisfactory when problems with the magnaflux procedure were discovered.

Because the engine was tested on a dynamometer rather than a propeller test stand, there was no air blast to cool the ignition harness. As a result, the harness was heat damaged and had to be replaced several times. Ultimately, harness-burning problems were overcome by fitting it with forced air-cooling.

One rocker support broke due to fatigue at a sharp angle at one of the inner flanges. More generous fillet radius prevented future occurrences of breakage at that point.

Following the test, wear measurements were made on all journals and bushings. No excessive wear was found. Most of the bushings were still within production limits. No bushing had been replaced during the test process because of excessive clearances.

Lessons Learned

Review of the T-14 test reports gives one the impression that the engine became of secondary importance after Wright Aero decided to pursue simultaneous development of the T-42. While work on the T-14 did indeed continue according to the original schedule, accounts of conferences and progress reports concentrated on the T-42. In a typical engine development program, individual components such as pistons, valves gear, and connecting rods would have been proven in single-cylinder testing. Once engineers were happy with the individual components, a larger engine, one the size of the T-14, would be built to prove the interaction of the individual components. Finally, the complete engine, in this case, the T-42, would be built and its problems sorted out. This seemingly lengthy process usually saves time and money. Wright Aero chose to skip single-cylinder testing altogether and proceed directly with the T-14. Then, before the T-14 was complete, the T-42 took precedence.

The formal reports on both the 50 and 150-hour tests of the T-14 indicate that no major weaknesses were encountered with the engine's unique features. Indeed, for an engine that was so immature, the T-14 tests went very well. However, careful study of Chapter 5, "The Tornado Program" and Appendix C —"Tornado 14 Tests" identifies a plethora of troublesome little details that later conspired against the T-42.

It is easy to look at the Tornado program and come up with ways to do it better. However, by putting oneself in the position of the engineers involved with the Tornado during its development, it is also easy to see how the natural optimism of engineers coupled with the pressures to demonstrate progress and lack of personnel could cause small indications of deeper issues to be ignored. The most difficult task that faces any chief engineer when a product enters the experimental phase is how to know which problems are important, and which ones are not.

Troubles with the T-14 fell into three general categories—assembly and operation errors, manufacturing and processing errors, and design shortcomings. Each will be discussed below.

Assembly and operation errors are expected, especially with a new engine. Since no one yet knows how to build it correctly, it sometimes is built incorrectly. With the T-14, hoses were left loose, oil plugs left out, spark plug wires crossed, seals damaged during installation, and parts incorrectly aligned. While these errors are regrettable, they are also to be expected. The same is true of operator errors, especially since the operators and test engineers were relatively junior personnel new to the program. They allowed the oil supply to run low, ran the engine at incorrect power settings, and failed to correctly diagnose problems caused by design, manufacturing, or assembly defects on a timely basis.

Manufacturing and processing errors contributed a share of trouble to the T-14. Process problems with the master rod bearings, incorrectly sized bushings, incorrectly hardened gears, and faulty magnetic particle inspections were all reported. Like assembly errors, these manufacturing errors were unfortunate, but expected. They were neither numerous nor severe and did no real damage to the Tornado program.

What did do damage to the Tornado program was a collection of design shortcomings that appeared in the T-14 but were either not noticed or ignored. To make matters worse, many of these design problems were the same design problems that nearly always gave trouble with any new engine. These included burned pistons, broken piston rings, broken cylinder hold-down studs, broken articulated rods, and worn vibration damper bushings. It is probable that many of these troubles would have been identified and fixed during single-cylinder testing.

Another almost universally troublesome engine component is the valve gear, and the Tornado was no exception. Stretched valves, burned valves, sticky valves, loose valve seats, broken rocker supports, and damaged rocker arms were all reported during T-14 testing. These problems later all caused grief with the T-42. It is obvious that no small amount of effort went into the solution of these valve problems, and likely that the Tornado valve troubles were close to being solved. Other engine programs with numerous and severe valve troubles had come to a happy end, and the Tornado should have not been an exception.

However, one final problem of major significance seemed to defy a solution. The T-14 was having trouble making power. If the T-42 was to be capable of 2,500 hp, then the T-14 had to produce more than 800 hp, but it was having trouble producing even 600 hp. A number of early experiments resulted in changes to valve timing and cam profiles. Later work involved reworking the intake tubes to give better flow from the supercharger diffuser. Neither approach was entirely successful.

Why Wright Aero did not pursue a solution to the power problem more aggressively is purely a matter of speculation. It is easy to imagine the engineering team reasoning that since the supercharger was not very well matched to the engine, the trouble must be there. It is also possible the team thought that since the larger T-42 would have a completely different manifold arrangement, effort making the T-14 manifolds more efficient would be wasted. Whatever the reason, it is clear that one essential element of the power problem was missed—pumping losses were higher than anticipated and were robbing the engine of power. Tornado power calculations had been made using BMEP values from Wright Aero's most powerful Cyclone, with huge cylinders and large valves running at relatively low speed. It simply was not possible to assign these same BMEP values to the high-speed Tornado with its many tiny cylinders.

---

[1] Contract No. W535 AC-12632.

[2] Swords, 50 Hour Development Test. See also Lundquist, 150 Hour Endurance Test.

[3] Ibid.

4 The Tornado 42

Three-quarter Front Right View of the Tornado 42 (Arthur Nutt Papers, Acc. 1987-0015, National Air and Space Museum, Smithsonian Institution, SI Neg. No. 2001-2657)

There were several important objectives in the Tornado 42 (T-42) design. Foremost, the designers wanted a lightweight engine that compactly arranged 42 cylinders, operating at high speed, and producing great power from a small diameter.[1] This was achieved by separate crankshaft assemblies for each group of 14 cylinders, physically connected through gears and torsionally resilient shafts. In this way, the crankshaft for a given cylinder group was only required to carry the torque generated by that group. This approach also isolated torsional vibration originating in one part of the engine from other parts of the engine, making the management of such vibration easier. Additionally, the designers hoped to advance the state of the art by producing an engine with improved cooling system, lubrication system, valve mechanism, propeller reduction gear, supercharger drive, and accessory drive.

The arrangement and construction of the T-42 included many innovations and unique internal details. Many of these have already been discussed in Chapter 3—"The Tornado 14." Unfortunately, few photographs of internal components and material specifications for the T-42 exist. Some additional information can be found in Appendix D—"Reprints of Tornado 42 Specifications" and Appendix E—"Data from Engine No. 11770."

The T-42 was composed of four main sections – the front section, the power section, the supercharger section, and the accessory section. A number of systems were also included. The sections and systems that made up the complete engine are discussed in the following paragraphs.

A view of the Tornado 42 left side shows the four main sections. (Courtesy of the New Jersey Aviation Hall of Fame and Museum)

Front Section

Figure 4.1 depicts the T-42 front section that consisted of propeller reduction gear components and torquemeter enclosed by the nose case.

The front end of the propeller shaft (71) was carried by a thrust bearing (73) and a radial bearing (74) at the front end of the nose section (67). A journal (72) at the rear end of the propeller shaft (71) was piloted in a bushing inside the front end of the front crankshaft.

Running on a journal near the rear end of the propeller shaft (71) was an intermediate power gear (76) having 91 teeth and an integral intermediate gear sleeve (75). This intermediate power gear (76) was driven by 21-tooth pinions (56) on each of the seven layshafts. At the front of the intermediate gear (76) were 18 inclined annular pockets (78) that were opposite similar inclined annular pockets (79) in a reduction driving gear (80). Within each of these pockets was a roller (81). These rollers (81) served to transfer driving effort from the intermediate gear (76) to the reduction driving gear (80), while at the same time creating an axial force between these two gears that was proportional to the torque transmitted. The reduction driving gear (80) was free to move a short distance axially along the intermediate gear sleeve (75). A cylinder (82) in the reduction driving gear (80) was engaged by a piston (83) that was rigidly attached to the intermediate gear sleeve (75).

Secured to the front crankcase diaphragm (55) was an oil transfer collar (70). Engine oil under pressure was admitted via passages in the oil transfer collar (70) and intermediate gear sleeve (75) to the cavity between the piston and cylinder. This pressurized oil counteracted the axial thrust caused by transmitted torque. The pressure of the oil in cylinder (82) was thus proportional to the torque transmitted. This pressure was registered on an appropriate gauge.

As pressure built in the cylinder (82), it caused the reduction driving gear to move aft relative to the piston (83), until the piston uncovered a bleed port (82') in the cylinder rim. This served to establish a pressure proportional to the torque transmitted. Oil from this bleed port flowed to a centrifugal balance cell (84) formed between the reduction driving gear (80) and a portion of the hub in the intermediate power gear (76). This cell was vented close to its most radially inward point so that centrifugally produced oil pressure in the cell (84) and in the cylinder (82) were balanced and did not affect the pressure proportional to torque in the cylinder (82).

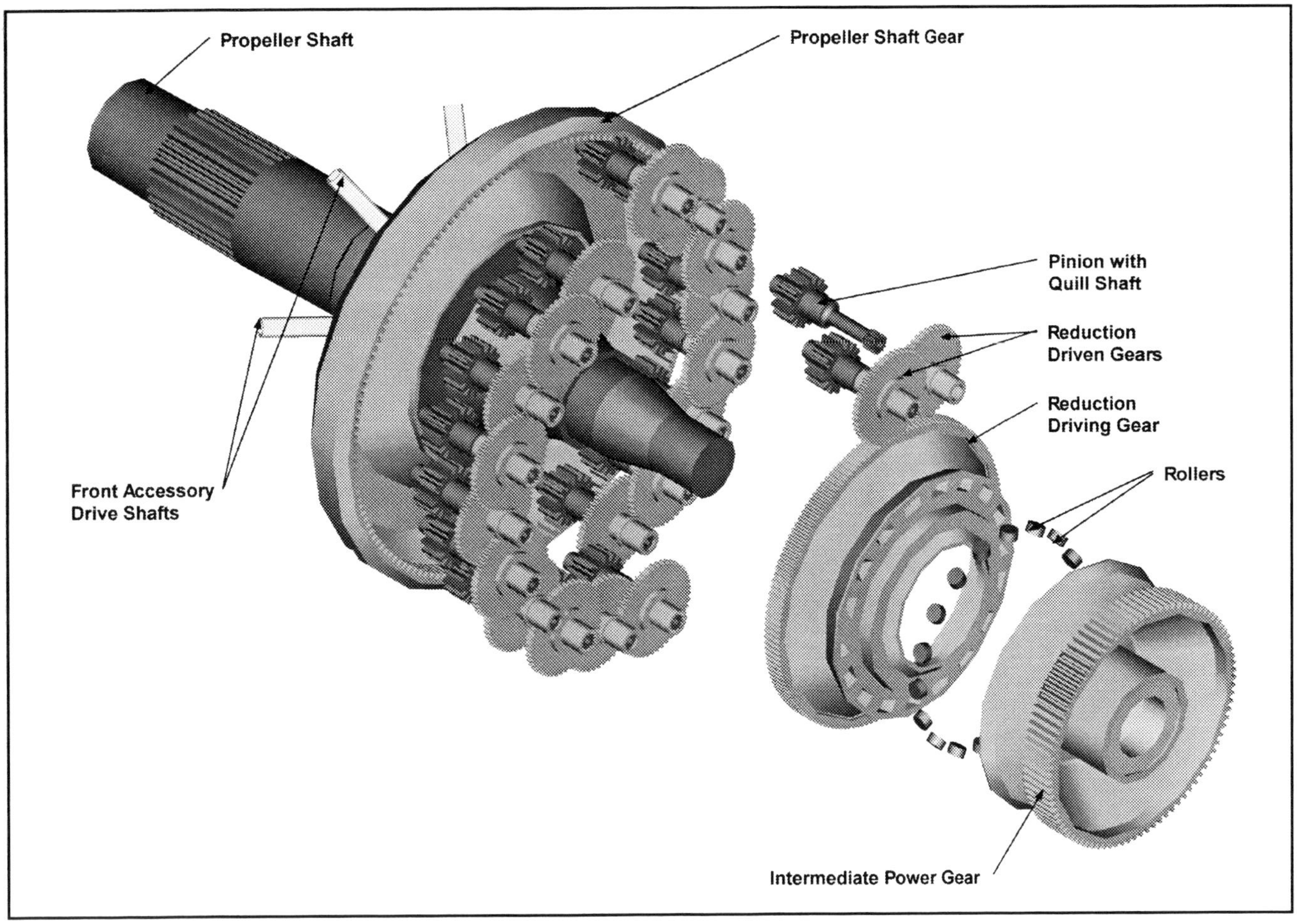

This exploded view of the propeller reduction gear shows the relative location of major components.

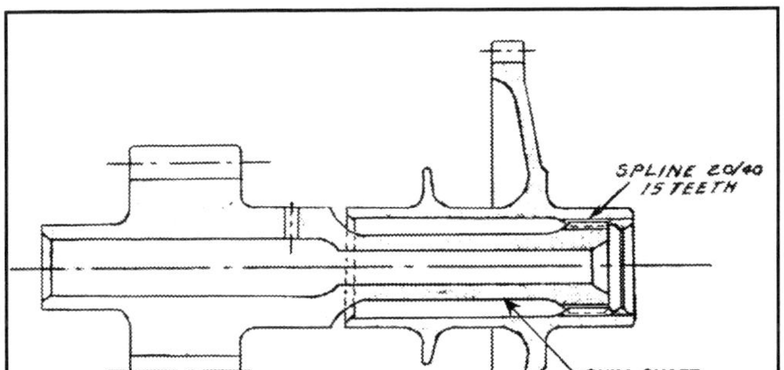

This cross section of the reduction driven gear and pinion shows the flexible quill shaft used to isolate crankshaft torsional vibration from the reduction gear and to ensure equal loading of all gears. A similar arrangement was used for the supercharger drive. (From Wright Field Technical Library Collection, Acc. XXXX-0428, National Air and Space Museum, Smithsonian Institution, Document D52.41 Wright/547, redrawn and annotated by the author)

Key to Figure 4.1

12 – Scavenge oil duct
18 – Front crankcase
42 – Cylinder cooling jacket
50 – Cylinder cover
51 – Front layshaft
55 – Front crankcase main bearing diaphragm
56 – Front layshaft integral drive pinion
57 – Front layshaft bearing
58 – Idler gear support diaphragm
59 – Front layshaft pinion
60 – Compound idler gear
61 – Idler gear
62 – Idler gear
63 – Camshaft gear
65 – Magneto
66 – Magneto drive shaft
67 – Crankcase nose section
68 – Fuel injector unit
70 – Torque meter oil transfer collar
71 – Rear end of propeller shaft
72 – Propeller shaft bearing journal
73 – Propeller shaft thrust bearing
74 – Propeller shaft radial bearing
75 – Intermediate gear sleeve
76 – Intermediate power gear
78 – Torque meter inclined annular pockets
79 – Torque meter inclined annular pockets
80 – Reduction driving gear
81 – Torque meter roller
82 – Torque meter hydraulic cylinder
82' – Torque meter bleed port
83 – Torque meter hydraulic piston
84 – Torque meter centrifugal balance cell
85 – Propeller reduction gear teeth
86 – Reduction driven gears
87 – Reduction driven gear shafts
88 – Bearing member
89 – Bearing member
90 – Bolts
91 – Reduction gear pinions
92 – Propeller shaft gear
93 – Spline connection
94 – Front oil annulus
95 – Oil pump driving bevel gear
96 – Front oil pump
97 – Propeller governor
117 – Annular oil groove
119 – Radial holes
120 – Oil passage
121 – Camshaft

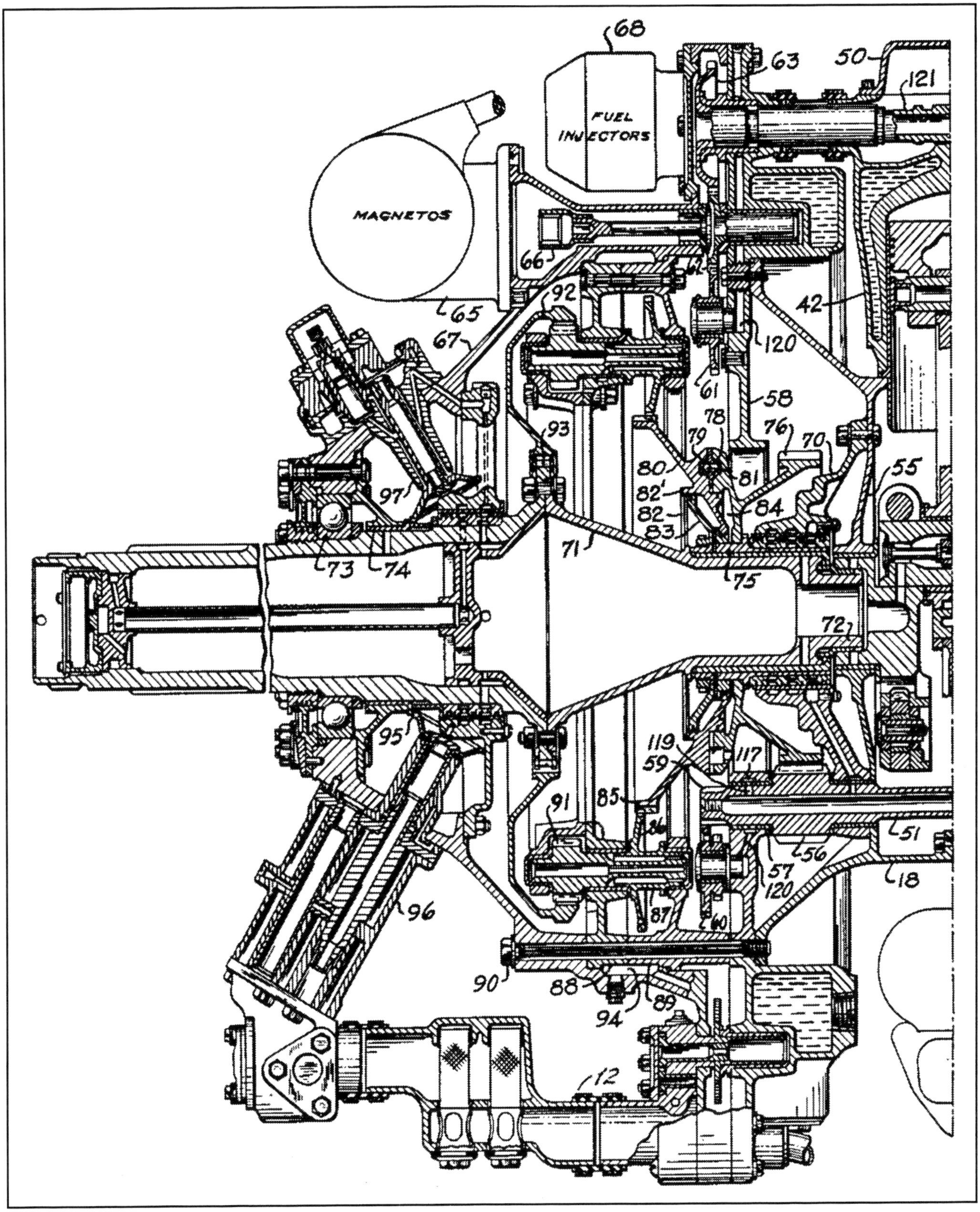

Figure 4.1. Longitudinal Section through the Engine Front Section (from U.S. Patent No. 2,355,277)

Key to Figure 4.2

16 – Short center crankcase
17 – Long center crankcase
20 – Crankcase section bolts
21 – Crankcase flanges
23 – Bearing support diaphragm
24 – Crankshaft center main bearing bushing
25 – Crankshaft center main bearing journal
26 – Two-throw crankshaft unit
27 – Crankshaft front main bearing journal
28 – Crankshaft front main bearing bushing
29 – Bearing support diaphragm
31 – Crankshaft rear journal
33 – Rear crankcheek
34 – Crankshaft drive gear
36 – Crankshaft unit front crankpin
37 – Crankshaft unit rear crankpin
38 – Crankshaft joint clamp bolt
40 – Cylinder base joint
41 – Combustion chamber
42 – Cylinder cooling jacket
43 – Cylinder liner
44 – Piston
45 – Master connecting rod
46 – Articulated connecting rods
48 – Valve cavity to crankcase passage
49 – Crankcase to cylinder bank passages
50 – Cylinder cover
51 – Layshaft
52 – Layshaft driven pinion sleeve
53 – Layshaft driven pinion bushing
54 – Layshaft driven pinion
121 – Camshaft
131 – Camshaft lower bearing half
132 – Camshaft bearing cap and rocker assembly
136 – Rocker arm roller

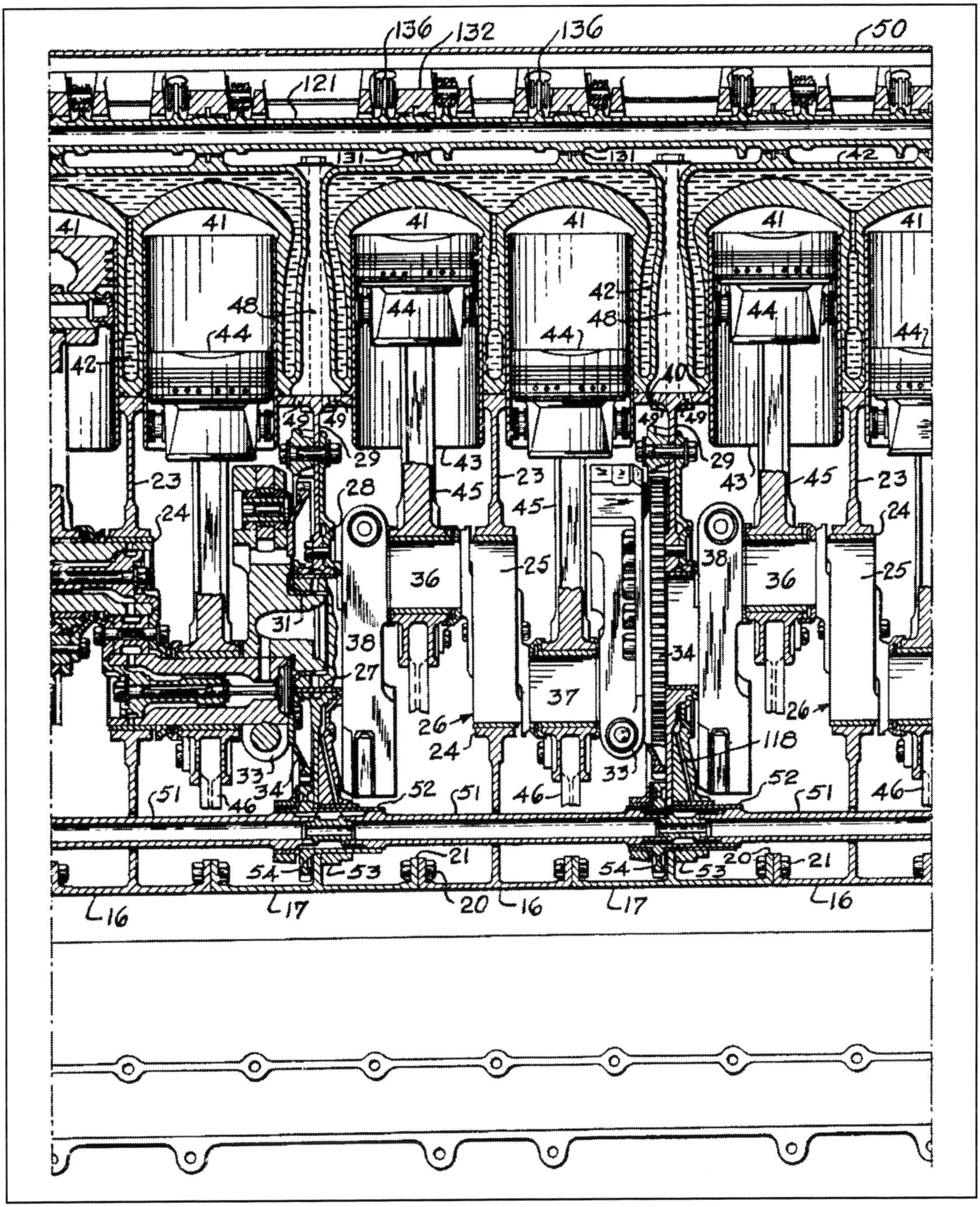

Figure 4.2. Longitudinal Section Through the Engine Power Section (from U.S. Patent No. 2,355,277)

Key to Figure 4.3

26 – Two-throw crankshaft unit
27 – Crankshaft front main bearing journal
28 – Crankshaft front main bearing bushing
29 – Bearing support diaphragm
31 – Crankshaft rear journal
33 – Rear crankcheek
34 – Crankshaft drive gear
37 – Crankshaft unit rear crankpin
38 – Crankshaft joint clamp bolt
40 – Cylinder base joint
41 – Combustion chamber
42 – Cylinder cooling jacket
43 – Cylinder liner
44 – Piston
45 – Master connecting rod
46 – Articulated connecting rods
48 – Passage from valve cavity to crankcase
49 – Crankcase to cylinder bank passages
51 – Layshaft
52 – Layshaft driven pinion sleeve
53 – Layshaft driven pinion bushing
54 – Layshaft driven pinion
118 – Drilled oil passages

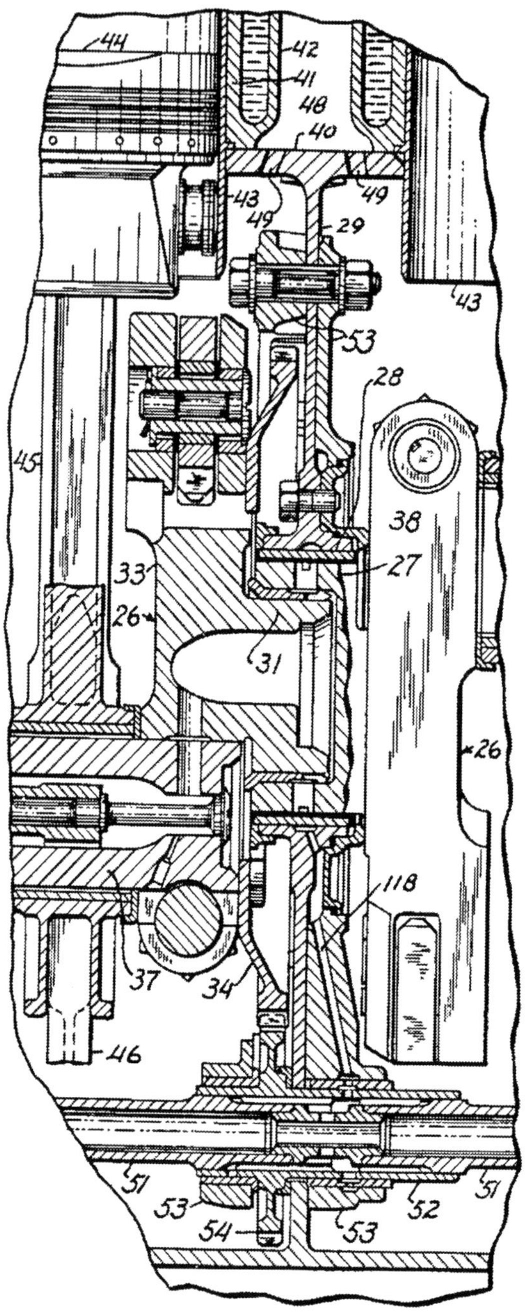

Figure 4.3. Enlarged Longitudinal Section through a Long Center Crankcase Bearing Support (from U.S. Patent No. 2,355,277)

Gear teeth (85) on the reduction driving gear (80) meshed with 45-tooth reduction driven gears (86) on shafts (87) captured between bearing members (88) and (89), which were in turn secured to the crankcase nose section (67) by bolts (90). Attached to each reduction driven gear via a short quill shaft was a 15-tooth pinion (91). The torsionally flexible quill shafts serve the dual purpose of helping to isolate crankshaft torsional vibration while ensuring equal loading for each of the 20 reduction gears. Meshed with the reduction driven gears was a 150-tooth propeller shaft gear (92) that was splined to the propeller shaft (71). The propeller shaft (71) was composed of rearward and forward portions, which were bolted together in the region of the spline connection (93). The bearing members (88) and (89), in conjunction with the crankcase nose section, define a front oil annulus (94), which will be discussed in more detail when the engine lubrication system is described later.

The propeller reduction gear components described above were for the 4.33:1 gear ratio nose case. It was sometimes desirable to change the reduction gear ratio, or to install a two-speed propeller drive or provisions for dual coaxial propellers. These changes were easily accomplished by altering the gears and elements (86 — 89, 91, and 92) as a bench assembly, separate from the engine, and installing this assembly as a unit to the remainder of the engine. This also allowed the torquemeter to remain intact with the engine. Since it was disposed between the engine and reduction gear, it always indicated true engine torque regardless of the reduction gear ratio.

Secured to the propeller shaft (71) immediately aft of the propeller shaft radial bearing (74) was a bevel gear (95) that drove a front oil pump (96) and a propeller governor (97). Both the oil pump and propeller governor were attached to the crankcase nose section (67).

Left Front View of the Propeller Reduction Gear Showing Front Accessory Drives

Power Section

The T-42 power section consisted of six polygonal steel crankcase units connected in tandem. The flat faces of these crankcase sections were aligned in a fore-and-aft direction, forming flat cylinder base joints to which seven cylinder banks of six cylinders each were secured. The cylinder banks were designated "A" through "G," starting at the top cylinder bank and rotating counterclockwise when viewing the engine from the anti-propeller end. The cylinder rows were designated "1" through "6," with "1" being the rearmost cylinder in each bank.

Three distinct two-throw crankshaft units were placed in tandem along the length of the engine. Each of these crankshaft units was geared to seven high-speed layshafts, disposed around the crankshafts and located between the skirts of the radially arranged cylinders. Each crankshaft unit was fitted with a large crankshaft drive gear with 84 teeth that drove layshaft driven pinions of 28 teeth at three times crankshaft speed. During assembly of the layshafts, the three crankshaft units were phased to one another at angles of 120°. This produced even firing of all forty-two cylinders.

Each row of cylinders included the usual pistons and connecting rods. Master connecting rods occupied the "A" bank, with articulated connecting rods in the remaining cylinders.

Each pair of crankcase units, along with its associated crankshaft unit, connecting rods, and pistons formed a power section module that was virtually identical to those previously discussed in Chapter 3. Cylinder banks were similar, but consisted of six-cylinder banks instead of the two-cylinder banks that had been used on the T-14.

Figures 4.2 and 4.3 depict the power section. While the T-14 crankcase consisted of only a front, center, and rear crankcase assembly, the T-42 crankcase consisted of a front crankcase (Figure 4.1 – 18) assembly, three short center crankcase (16) assemblies, two long center crankcase (17) assemblies, and a rear crankcase (Figure 4.3 – 19) assembly. The short center crankcase (16) assemblies were very much like those from the T-14. The long center crankcase (17) assemblies provided

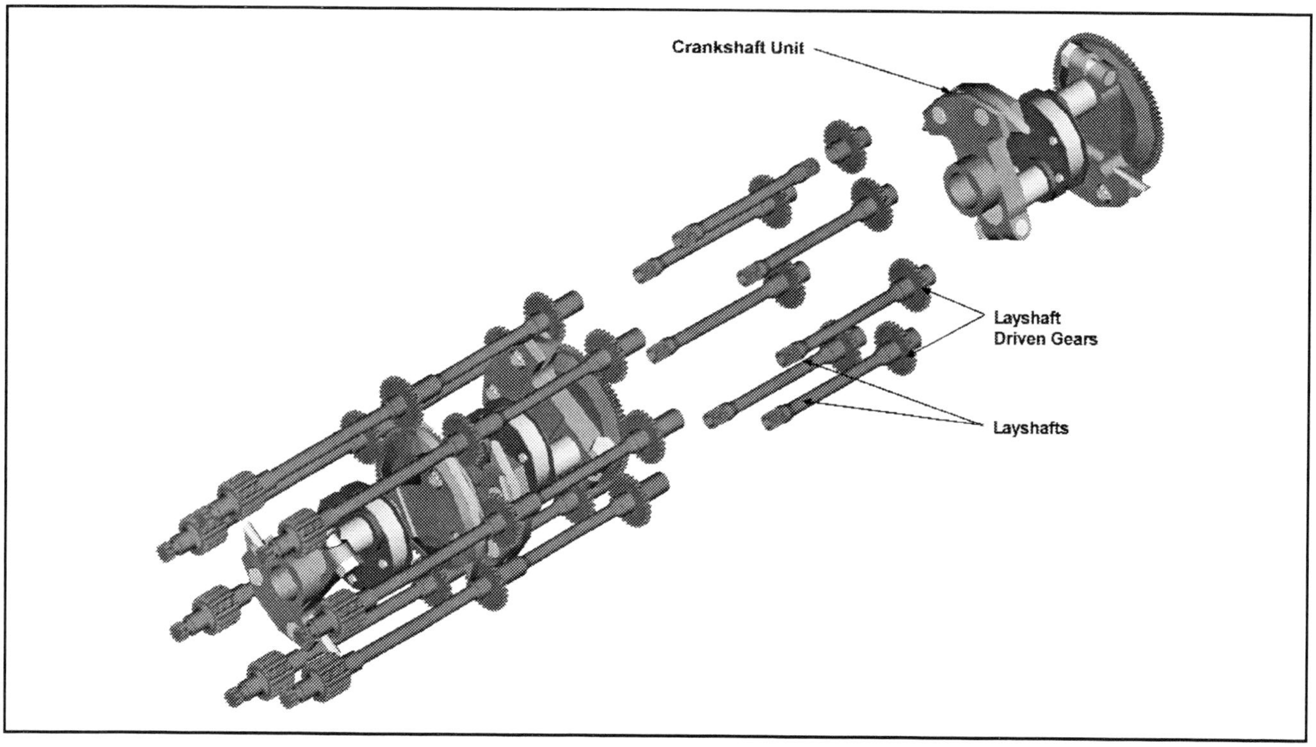

Three two-throw crankshaft units were geared to seven high-speed layshafts.

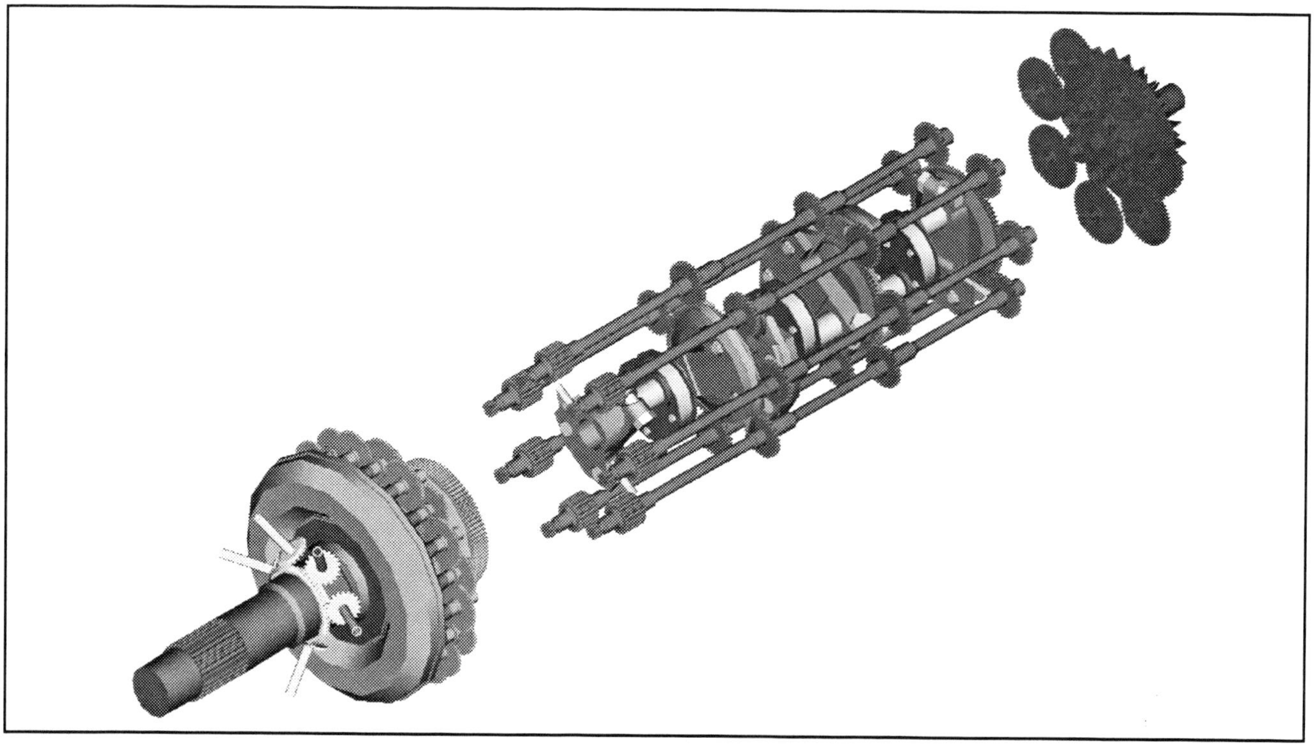

The completed power train drove the propeller reduction gear (left) and the supercharger gear train (right). Torsionally flexible layshafts ensured equal gear tooth loading.

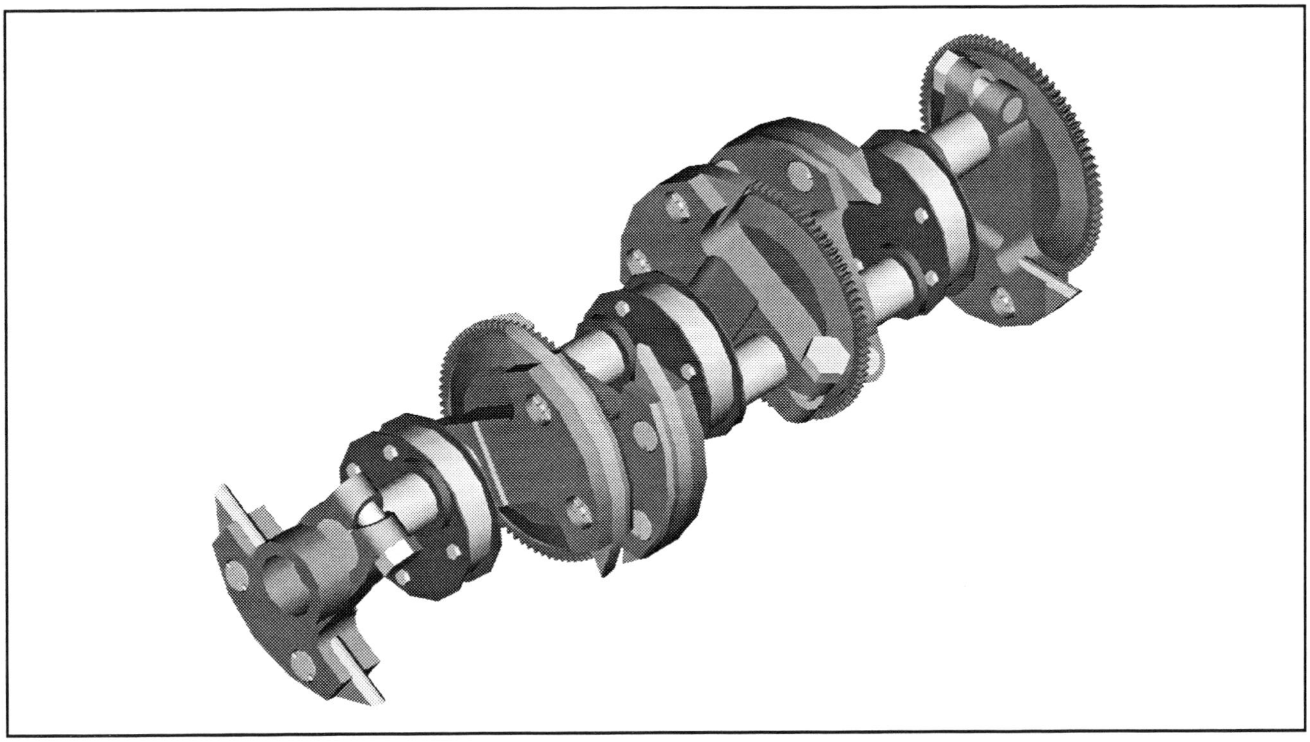

The crankshaft units were phased at 120 degrees to give even firing intervals for all 42 cylinders

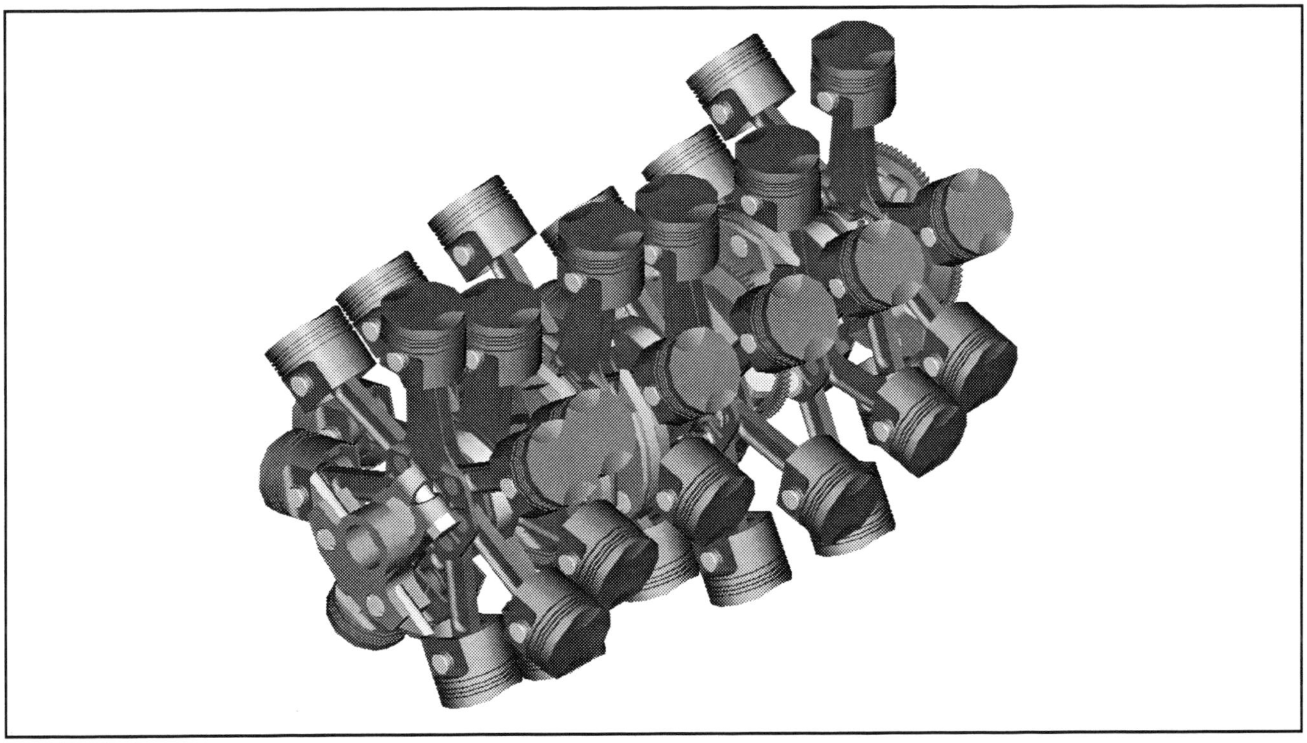

View of the Crankshaft with all 42 Connecting Rods and Pistons Installed

Key to Figure 4.4

19 – Rear crankcase
34 – Crankshaft drive gear
42 – Cylinder cooling jacket
50 – Cylinder cover
51 – Layshaft
52 – Layshaft driven pinion sleeve
53 – Layshaft driven pinion bushing
54 – Layshaft driven pinion
98 – Layshaft pinion bearing member
99 – Layshaft supercharger drive extension shaft
100 – Layshaft supercharger drive gear
101 – Rear crankcase bearing member
102 – Front supercharger housing
103 – Front supercharger housing bolt
104 – Supercharger drive idler gear
105 – Supercharger impeller shaft gear
106 – Supercharger impeller shaft
107 – Supercharger impeller
108 – Rear crankshaft extension
109 – Rear crankshaft spline
110 – Starter drive dog
111 – Crankcase accessory
112 – Accessory drive gear
113 – Accessory driven gears
114 – Accessory drive shafts
115 – Rear oil pump
116 – Rear oil annulus
122 – Accessory drive gear
130 – Accessory drive low-pressure oil annulus
142 – Intake manifold
148 – Supercharger diffuser chamber
150 – Supercharger inlet duct
156 – Rear annular coolant manifold
157 – Rear coolant conduits

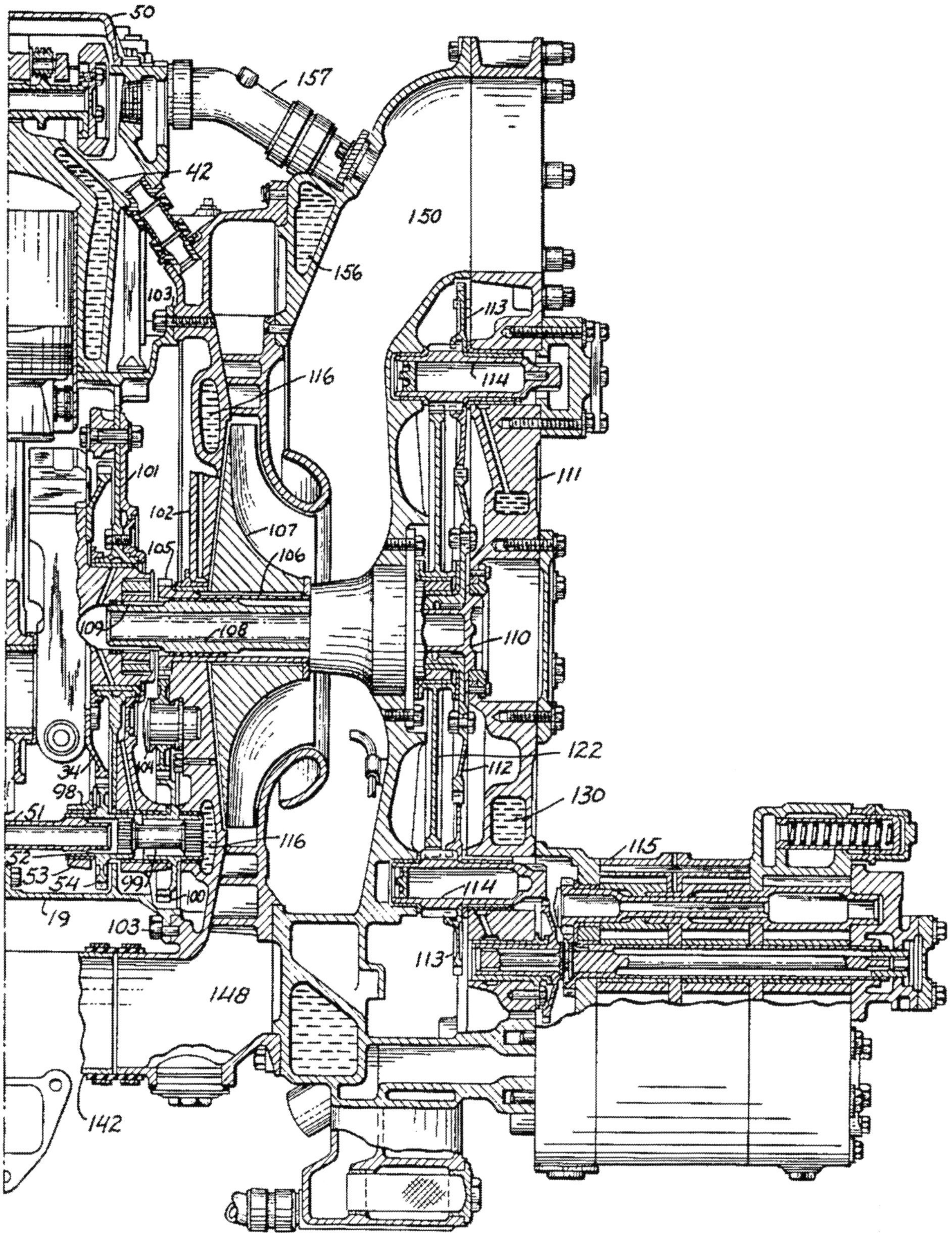

Figure 4.4. Longitudinal Section Through the Engine Supercharger and Accessory Sections (from U.S. Patent No. 2,355,277

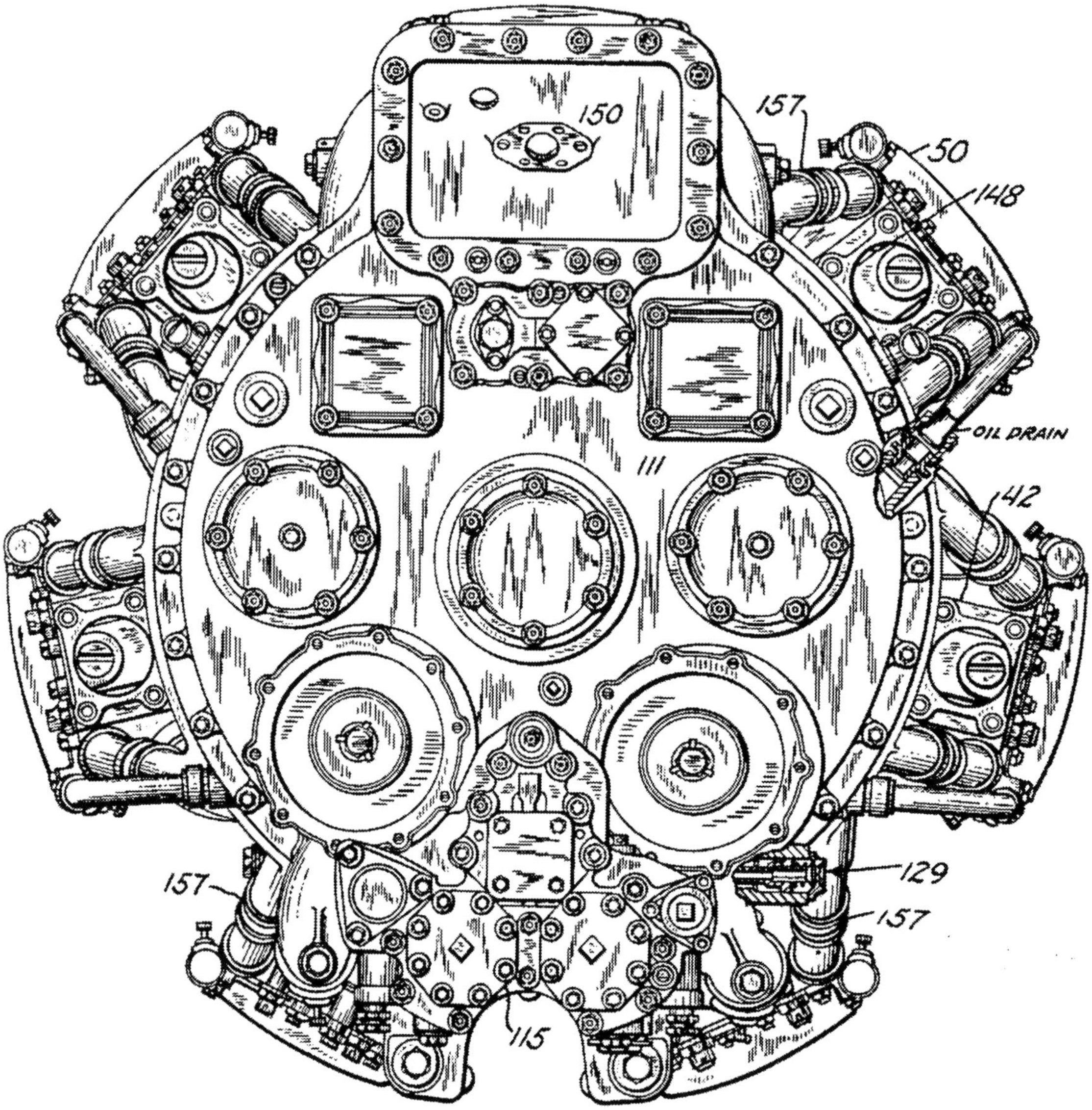

Figure 4.5. Rear Elevation of the Engine (from U.S. Patent No. 2,355,277)

Key to Figure 4.5

42 – Cylinder cooling jacket
50 – Cylinder cover
115 – Rear oil pump
129 – Oil pressure reducing valve
148 – Supercharger diffuser chamber
157 – Rear coolant conduits

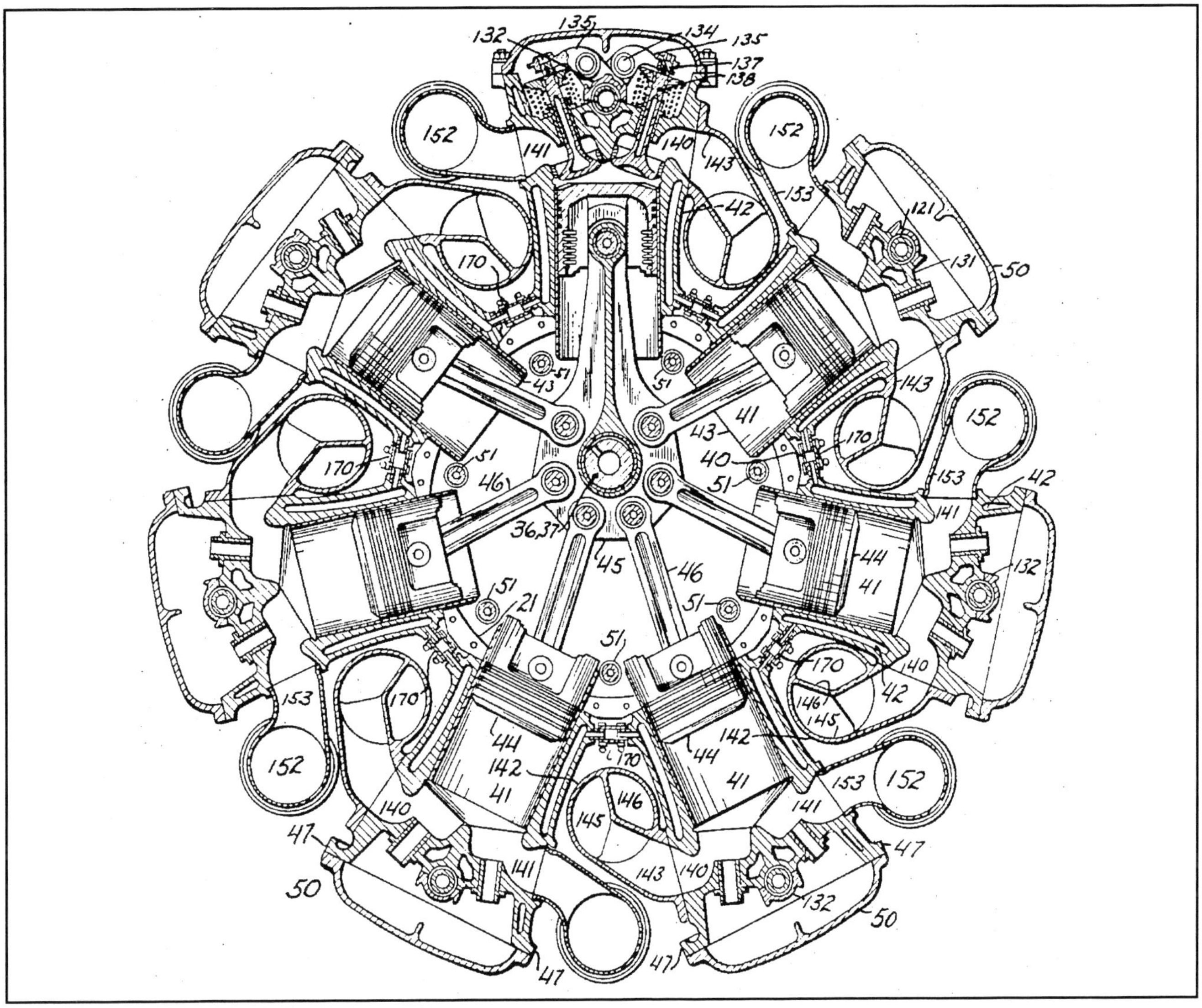

Figure 4.6. Section through a Cylinder Row (from U.S. Patent No. 2,355,277)

Key to Figure 4.6

21 – Crankcase flanges
36 – Crankshaft unit front crankpin
37 – Crankshaft unit rear crankpin
41 – Combustion chamber
42 – Cylinder cooling jacket
43 – Cylinder liner
44 – Piston
45 – Master connecting rod
46 – Articulated connecting rods
47 – Cylinder cover mounting flange
50 – Cylinder cover
51 – Layshaft
121 – Camshaft
131 – Camshaft lower bearing half
132 – Camshaft bearing cap and rocker assembly
134 – Rocker arm bearings
135 – Rocker arms
137 – Valve clearance adjustment
138 – Valve stems
140 – Intake ports
141 – Exhaust ports
142 – Intake manifold
143 – Intake conduit
145 – Intake manifold passage
146 – Intake manifold passage
152 – Exhaust manifold
153 – Exhaust conduit
170 – Cylinder row coolant connectors

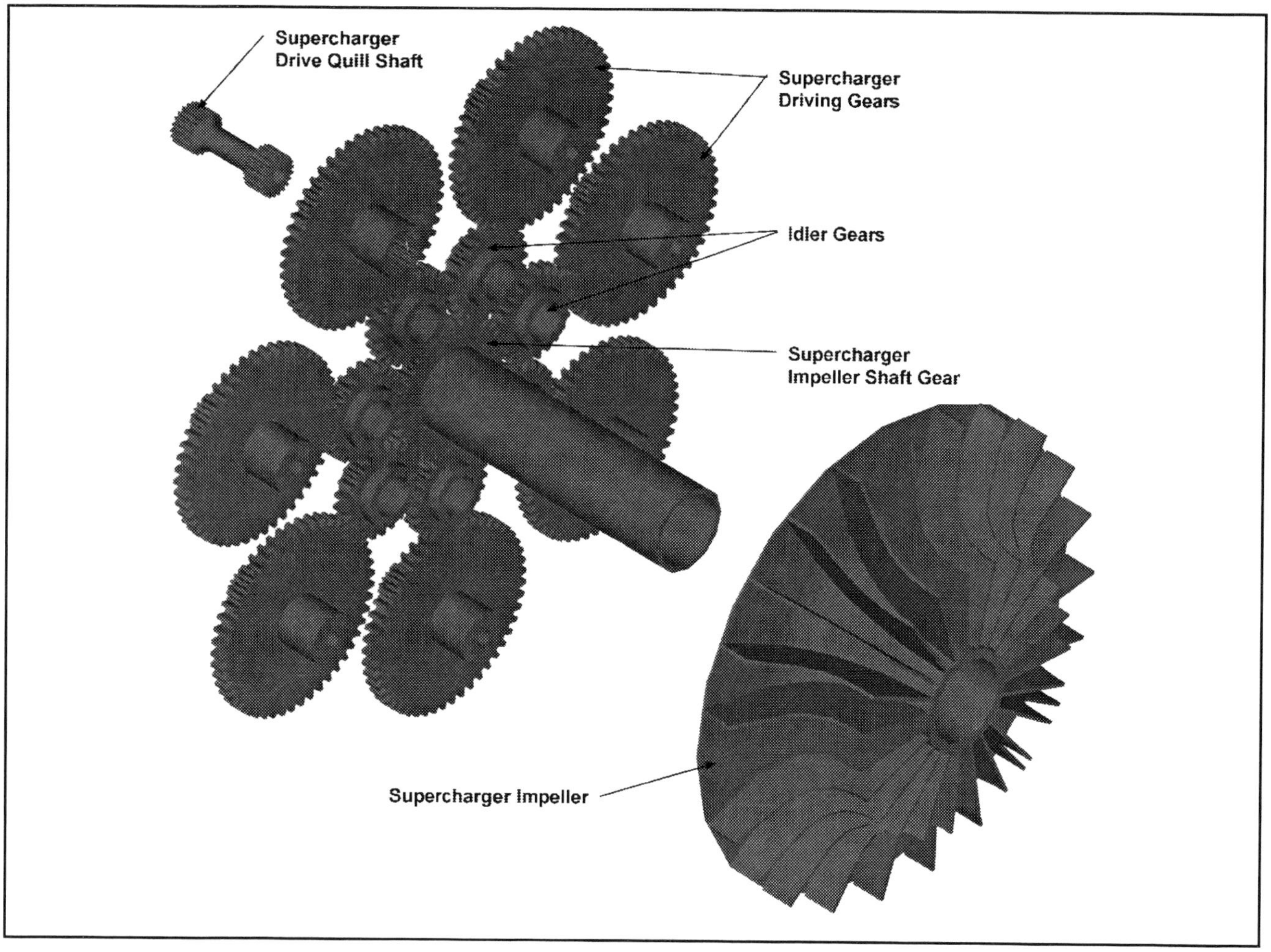

Exploded View of Supercharger Drive and Impeller

additional length for the crankshaft drive gears (34), layshaft driven pinions (54), layshaft driven pinion bushing (53), valve cavity to crankcase passage (48), and cylinderdeck passages (49). Eighty-four crankcase section bolts (20), installed in pairs through crankcase flanges (21) between each cylinder, held the crankcase together.

Supercharger and Accessory Sections

Figures 4.4 and 4.5 depict the supercharger and accessory sections, along with the associated supercharger impeller and accessory drive trains. The rear crankcase (19) carried a bearing member (98) in which ran the seven rearmost layshaft driven pinions (54). Each pinion (54) had a spline connection with the rear layshaft (51) The layshaft driven pinion sleeve (52) of each pinion was splined to a short supercharger drive quill shaft (99) that was also splined to a 49-tooth* supercharger driving gear (100). This gear ran in bearings within a rear crankcase bearing member (101) and a front supercharger housing member (102), the latter being secured by front supercharger housing bolts (103) to the rear crankcase (19). Torsionally flexible quill shafts isolated the supercharger drive train from crankshaft torsional vibration and ensured equal loading for all seven drive trains. Idler gears (104), having 24 teeth, were carried by the front supercharger housing member (102) and were driven by

*When the CAD model of the Tornado was being prepared, it became evident that the original longitudinal cross section drawings were in error. The intermediate supercharger drive gears were far too large and interfered with one another. Details of the supercharger drive train were reverse engineered by the author and represent a best guess as to individual gear sizes and number of teeth.

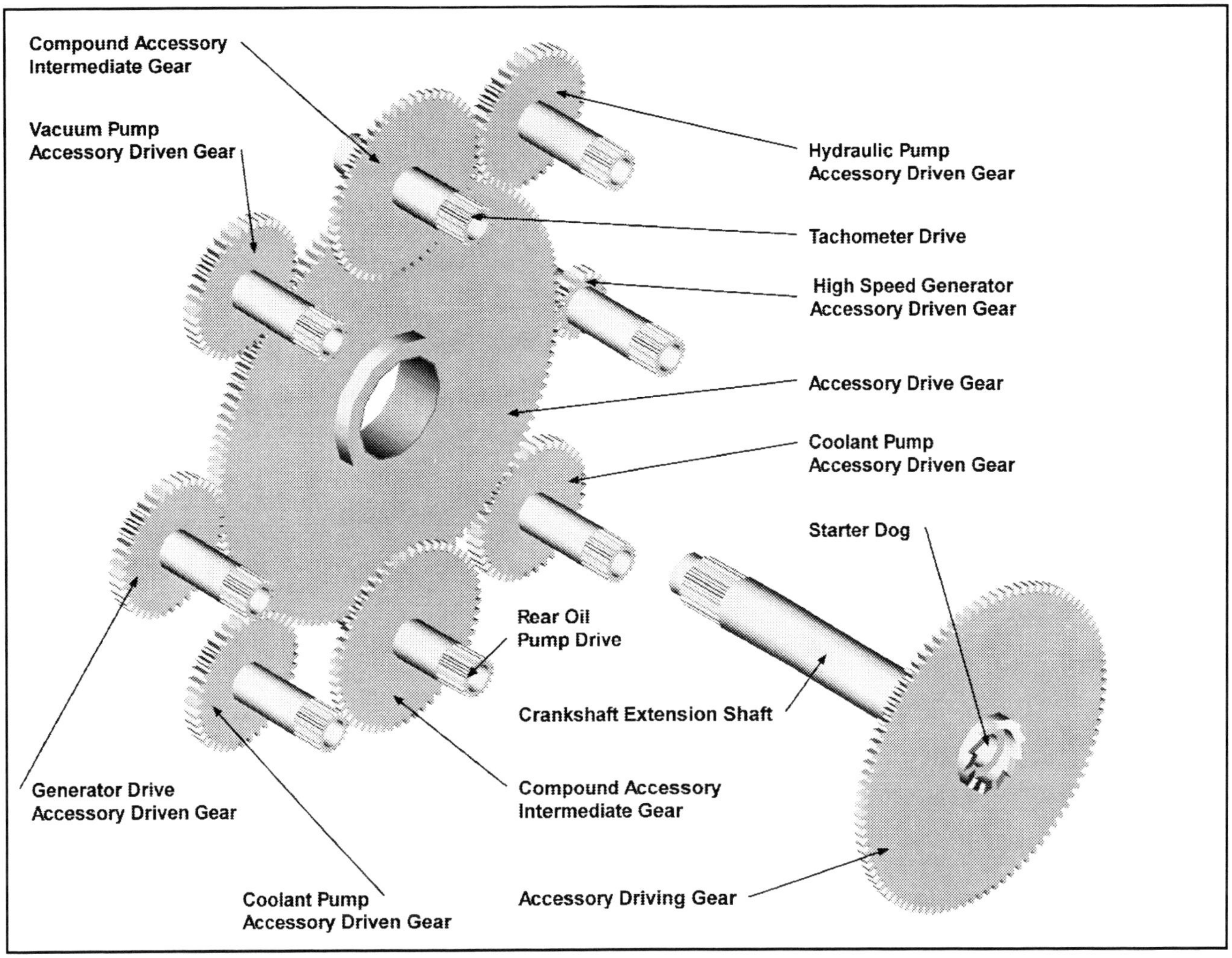

Exploded View of the Rear Accessory Drive. The spline at the front of the crankshaft extension shaft mated with the rear crankshaft unit. The fuel pump was mounted to the extreme rear end of the rear oil pump.

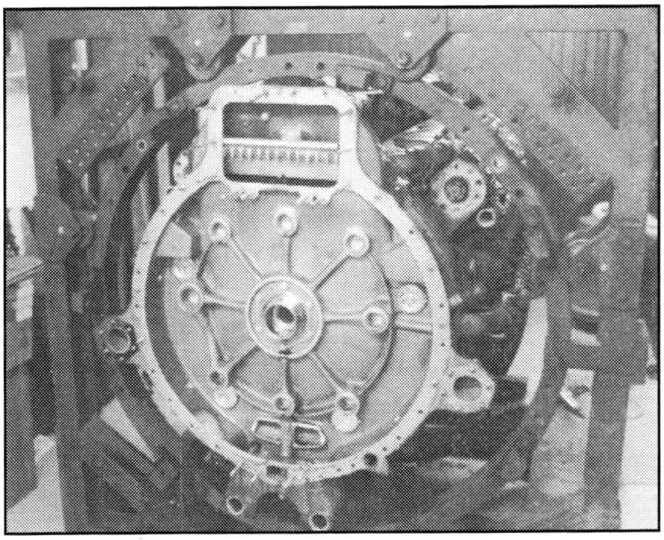

View Looking Forward into the Front Half of the Accessory Housing. Note eight bushings for the accessory driven gears and the supercharger inlet duct. (Courtesy of the David S. Way)

the supercharger driving gears (100). The idler gears drove the central 35-tooth supercharger impeller shaft gear (105) and integral impeller shaft (106), resulting in a gear ratio of 4.20:1. The impeller shaft was splined to the supercharger impeller (107). The supercharger impeller shaft (106) was piloted upon a rear crankshaft extension (108) whose front end was splined to the rear crankshaft spline (109). The rear end of the crankshaft extension shaft (108) was splined to a starter drive dog (110) secured in the crankcase accessory section (111).

Accessory Drive

A 96-tooth accessory driving gear (112), fastened to the rear end of the crankshaft extension shaft (108), drove two compound accessory intermediate gears (113) and associated accessory drive shafts (114). The large gear of the accessory intermediate gear had 54 teeth, and the small gear had 18 teeth. The accessory drive shafts (114) provided power takeoffs for tachometer drives, fuel pump, and oil pumps (115). The small gear of the accessory intermediate gear drives a 132-tooth accessory drive gear (122). This, in turn, drives six accessory driven gears (not shown in Figure 4.3), which provide power to a low-speed generator, a high-speed generator, two coolant pumps, a hydraulic pump, and a vacuum pump. All of these accessory driven gears have 40 teeth and provide a drive ratio of 0.8 crankshaft speed, with the exception of the high-speed generator drive gear, which has 18 teeth and provides a drive ratio of 1.775:1.

The crankshaft extension shaft (108) was at the same end of the rear crankshaft as its drive gear (34), ensuring smooth torque to the accessory drives, uninfluenced by torsional windup and vibration in the crankshaft. In many other engines, power was taken from one end of the crankshaft while accessories were driven from the other, unbalancing the phase relation between the two ends of the crankshaft.

Direction of rotation could be reversed by changing the camshafts, supercharger impeller and diffuser, and altering the drive gearing in the accessory section to allow accessories to rotate in the same direction as before the change was made. This accessory section change would have been accomplished by adding idler gears between the accessory drive gear (112) and accessory driven gears (113).

Detail of the rear accessory housing and accessory drive gear train (Courtesy of David S. Way)

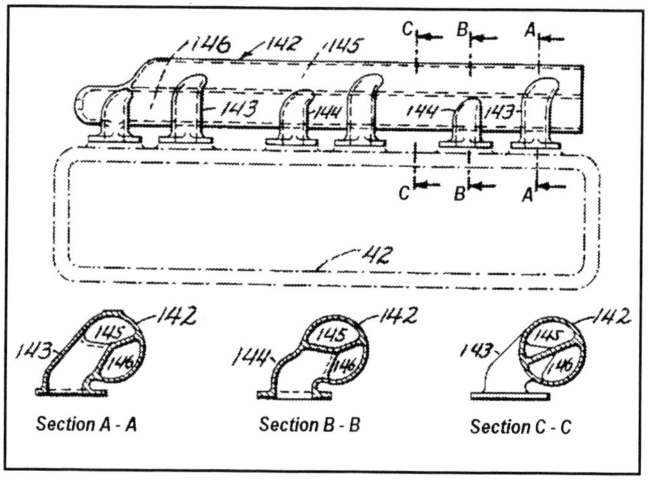

Figure 4.7. Intake manifold arrangement (from U.S. Patent No. 2,355,277)

Key to Figure 4.7

142 – Intake manifold
143 – Intake conduit
144 – Intake conduit
145 – Intake manifold passage
146 – Intake manifold passage

Induction System

As shown in Figure 4.6, the clockwise side of each cylinder bank was provided with intake ports (140), while the counterclockwise side of each cylinder bank was provided with exhaust ports (141). An intake manifold (142) (see also Figure 4.7) was situated in the vee between each bank of cylinders and was connected to the intake ports (140) by intake conduits (143 and 144). The intake manifold (142) was formed with parallel separate internal passages (145 and 146). The intake conduits (143) connected the alternate odd-numbered cylinders of the bank with intake manifold passage (145). The intake conduits (144) connected the alternate even-numbered cylinders of the bank with intake manifold passage (146). The rear end of each intake manifold (142), as shown in Figure 4.4, was connected to the diffuser chamber (148) through which the fuel/air mixture from the supercharger impeller (107) flowed. A horizontally mounted carburetor (not shown) was connected to the face of the crankcase accessory section (111) and delivered a fuel/air mixture to the supercharger impeller (107) via the supercharger inlet duct (150). If fuel injection was provided, only air was admitted to the supercharger.

The problem of uniform air or mixture distribution to an engine having a large number of cylinders was a difficult one. The double passage manifolds, described above, constitute a means of assuring uniform distribution, by dividing the flow into 14 different passages that, in effect, provide a separate manifold for each group of three cylinders. The selection of cylinders fed by each manifold is based upon a long interval in the firing order of the engine so that overlap in flow to more than one cylinder of each manifold, at any one time, is avoided.

Figure 4.6 shows that an exhaust manifold (152) is situated between each bank of cylinders radially outward of the intake manifold and is connected to the exhaust ports (141) by exhaust conduits (153).

Above: A view between cylinder banks with the intake manifold removed. (Courtesy of David S. Way)

Left: This view between two cylinder banks shows the intake manifold, associated fasteners, and safety wiring. Exhaust ports on the opposite bank were plugged. (Courtesy of David S. Way)

Overhead Camshaft and Valve Gear

Figures 4.2, 4.6, and 4.8 show that each camshaft (121) extended the full length of each cylinder bank and ran in a lower bearing half (131) formed in the head of each cylinder and carrying a suitable half bushing. Camshaft bearing cap and rocker assemblies (132), which secured the camshaft to the bearing half (131), were bolted to the cylinder head and included two rocker shafts (134) axially parallel to the camshaft and spaced above and to either side of the camshaft. These rocker shafts carried rocker arms (135) with rollers (136) that engaged the camshaft, and valve clearance adjusting screws (137) that engaged the tops of valve stems (138). This combination of bearing cap and rocker carrier avoided the use of many small parts and afforded easy assembly and disassembly of the rocker and camshaft assembly. There was one cap and rocker assembly (132) for each cylinder. A camshaft bearing between each pair of cams and rocker arms minimized bending deflections in the camshaft. The valves (138) ran in valve guides and interfaced with valve seats inserted in a conventional manner, as shown in Figure 4.6. The camshafts were driven by the layshafts through a gear train as shown in Figures 4.9 and 4.10.

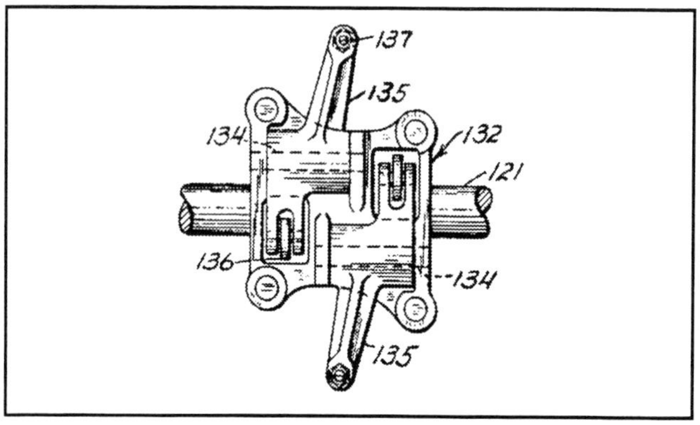

Figure 4.8. Rocker arm and camshaft bearing bracket (from U.S. Patent No. 2,355,277)

Key to Figure 4.8

121 – Camshaft
132 – Camshaft bearing cap and rocker assembly
134 – Rocker arm bearings
135 – Rocker arms
136 – Rocker arm roller
137 – Valve clearance adjustment

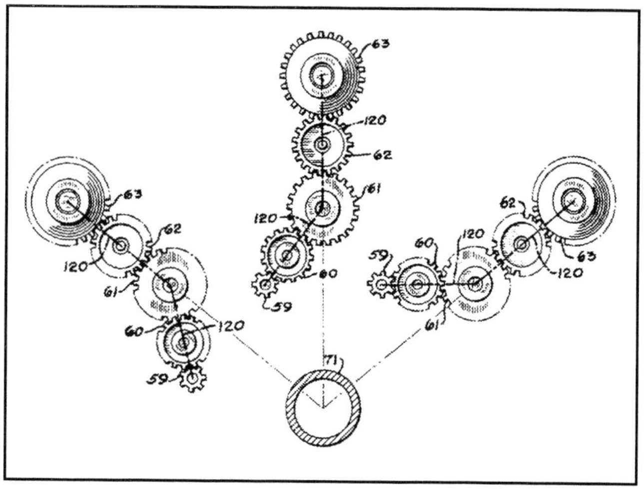

Figure 4.9. Camshaft drive train and lubrication (from U.S. Patent No. 2,511,248)

Key to Figure 4.9

59 – Front layshaft pinion
60 – Compound idler gear
61 – #2 Idler gear
62 – #3 Idler gear
63 – Camshaft gear
71 – Rear end of propeller shaft
120 – Oil passage

Detail of camshaft, camshaft supports, rocker arms, and valve adjustment provisions (Courtesy of the New Jersey Aviation Hall of Fame and Museum)

Lubrication System

The T-42 lubrication system was of the dry sump type. Oil from an oil tank was presented to the inlet of oil pressure pumps, which forced the oil through various oil passages within the pressure circuit of the engine. Once it had performed its lubrication function, the oil was collected, or scavenged, and returned to the oil tank by scavenge pumps.

Pressure Circuit

Figure 4.11 depicts the lubrication system pressure circuit. The front and rear oil pumps (96 and 115) included both a pressure pump and a scavenging pump. The pressure pumps delivered oil through suitable conduits in the crankcase sections to a front oil annulus (94) and (Figure 4.1 – 94), and a rear oil annulus (116) and (Figure 4.4 – 116) integral with the front supercharger housing (Figure 4.4 – 102). The rear oil annulus (116) passes across the rear end of each layshaft (51). The layshafts were hollow, allowing oil under pressure to pass from the rear oil annulus forward along the layshafts. At appropriate intervals, openings in the layshafts

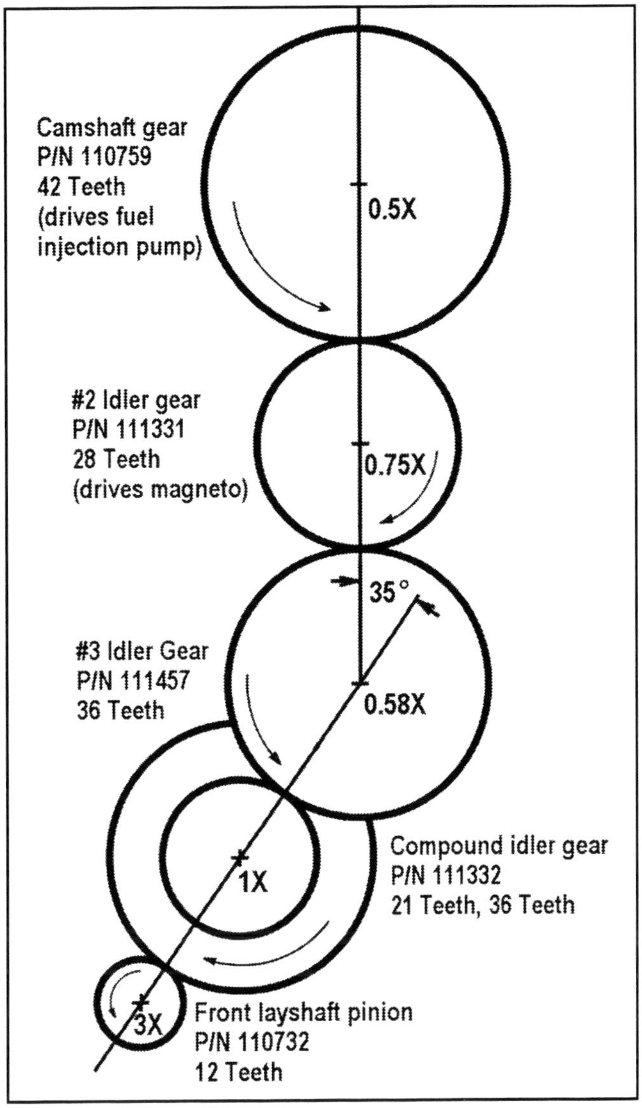

Left: Figure 4.10. Schematic of the "A" bank camshaft drive train. The other six camshaft drive trains were identical. The "X" on each gear refers to its rotation speed relative to the crankshaft.

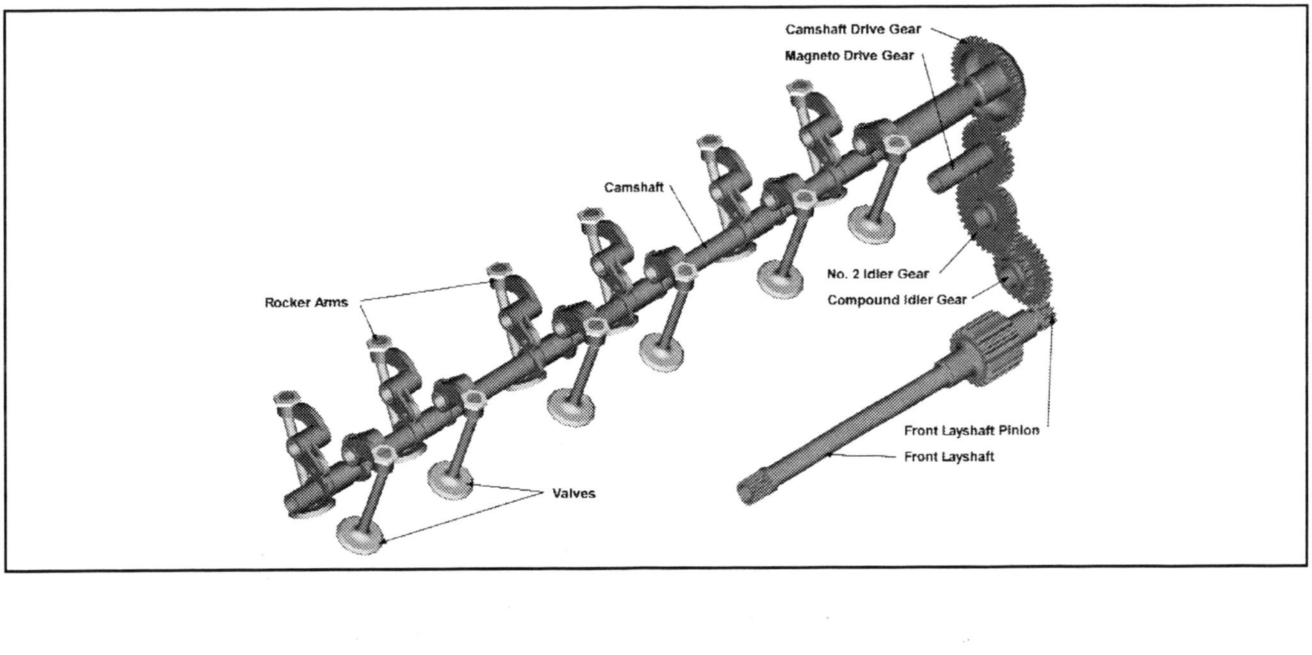

Below: Detail of Valve Gear Train, Camshaft, Rocker Arms, and Valves

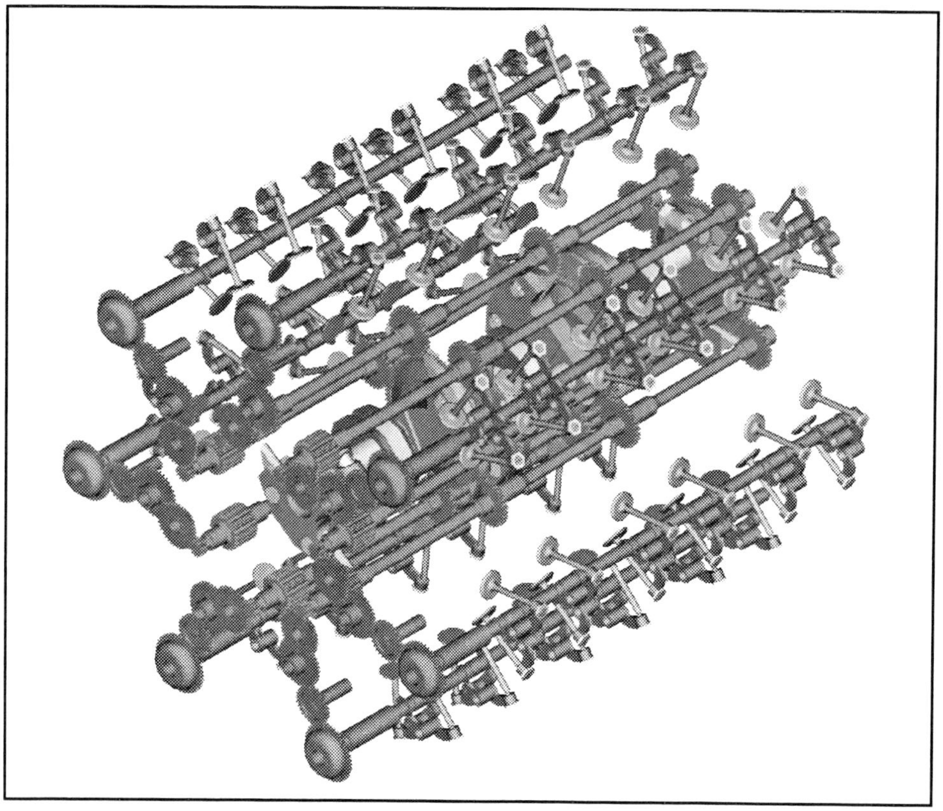

A View of the Rotating Power Section Parts with Valve Gear Included

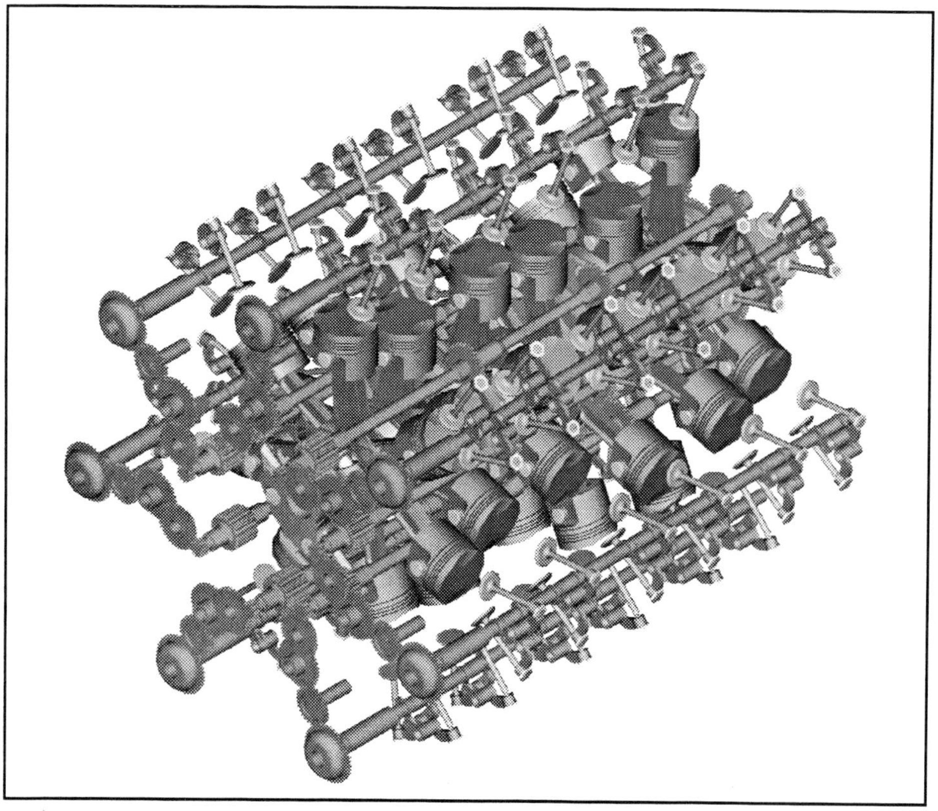

The Addition of Reciprocating Parts Completes the Power Section

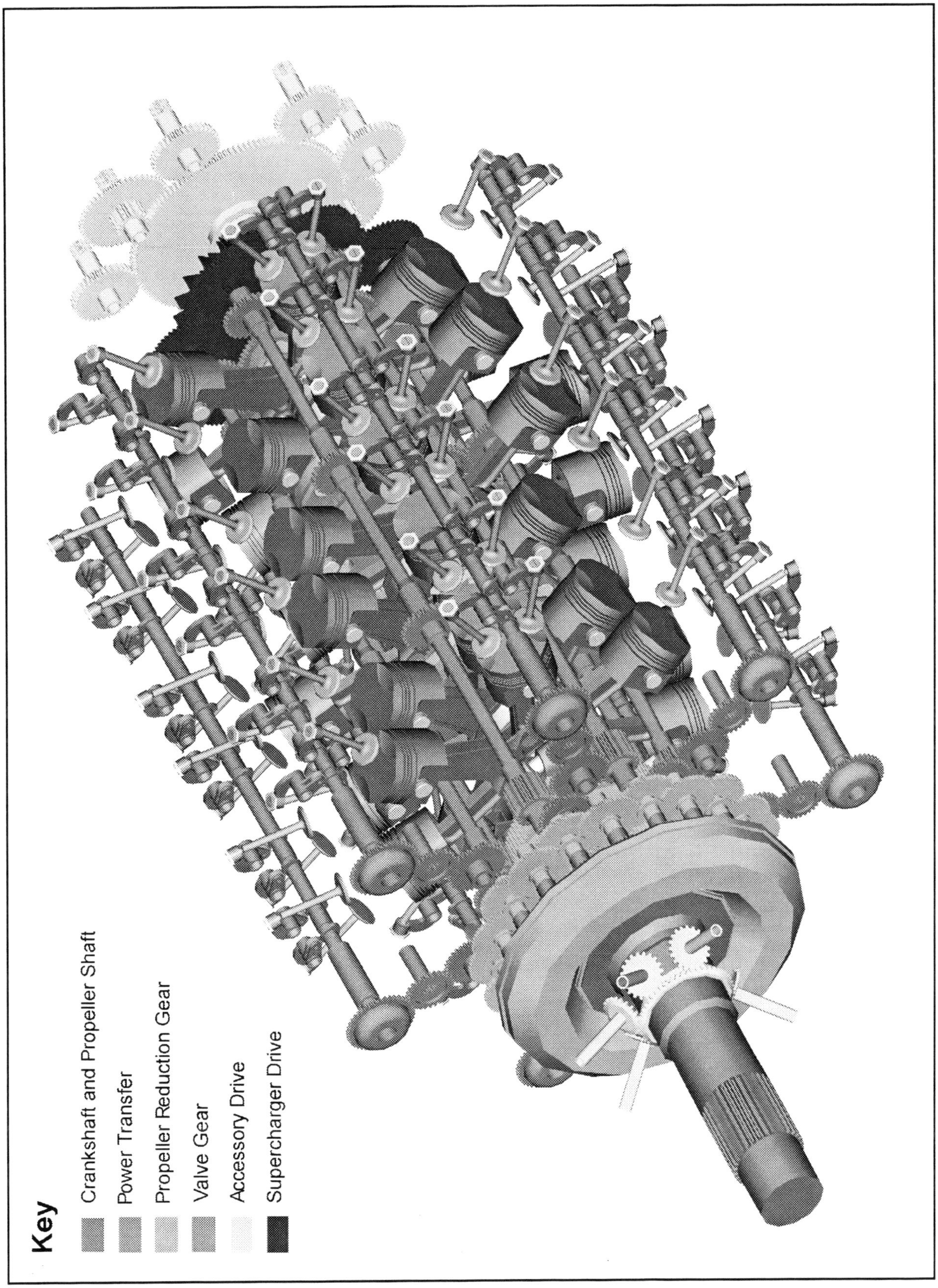

The complete engine is composed of the power section, propeller reduction gear, supercharger, and accessory drive.

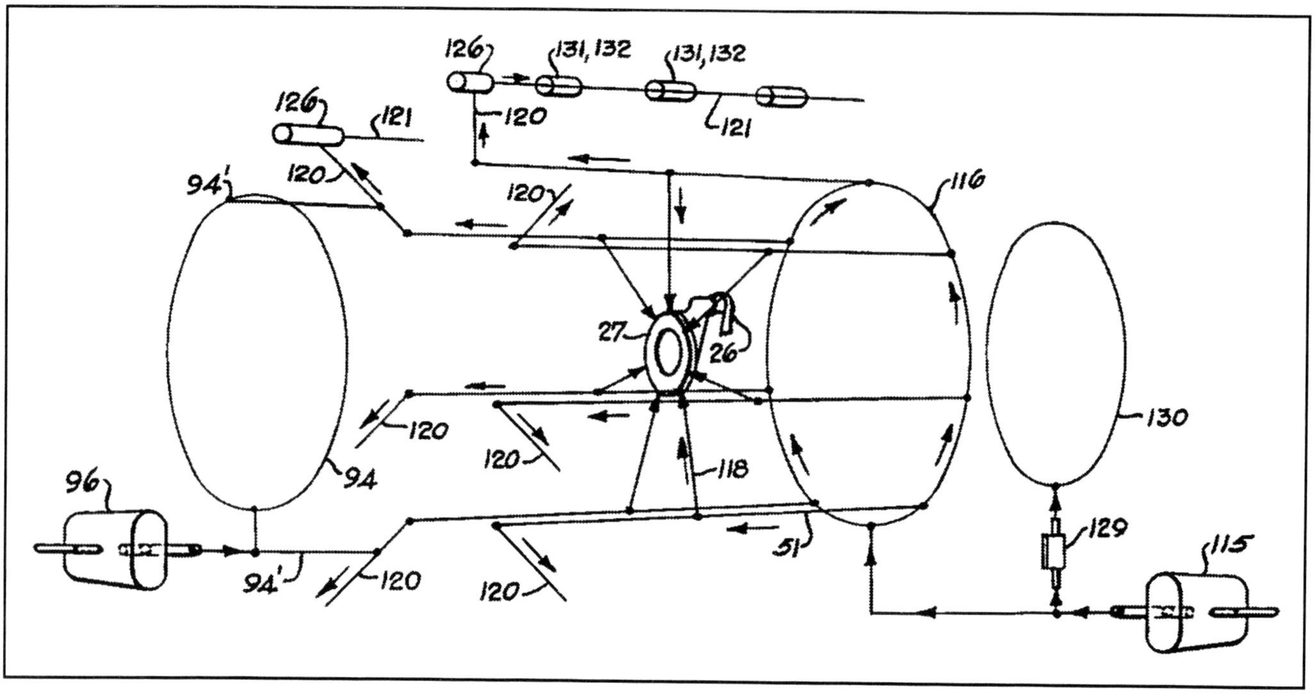

Figure 4.11. Schematic of the Engine Lubrication system (from U.S. Patent No. 2,522,248 with additions by the author)

Key to Figure 4.11

26 – Two-throw crankshaft unit
27 – Crankshaft front main bearing journal
51 – Layshaft
94 – Front oil annulus
94' – Oil passages
96 – Front oil pump
115 – Rear oil pump
116 – Rear oil annulus
118 – Drilled oil passages
120 – Oil passages
121 – Camshaft
126 – Camshaft oil pressure reducer valve
129 – Oil pressure reducing valve
130 – Accessory drive low-pressure oil annulus
131 – Camshaft lower bearing half
132 – Camshaft bearing cap and rocker assembly

communicated with holes in bearing bushings which admitted oil into oil passages such as (118) which led to the crankshaft main bearings. The crankshaft main bearings were drilled to conduct oil to the crankshaft interiors and to the crankpin bearings for pressure lubrication of the connecting rod bearings and knuckle pin bearings. The cylinder walls, piston pins and pistons were lubricated by oil thrown from the master rod and knuckle pin bearings.

The front layshafts (51), shown in Figure 4.1, were provided with radial drillings (119) that communicated with an annular groove (117) in each front layshaft bearing (57). These grooves connected with seven oil passages (120) that were, in turn, connected to the front oil annulus (94) via oil passages (not shown in Figure 4.1, but schematically depicted in Figure 4.11 as 94'). In this manner, oil under pressure was pumped by both the front and rear oil pumps to each end of the layshafts, providing a redundant supply of high pressure oil to the entire lubrication system.

Oil from the front oil annulus (94) also served to lubricate other sundry bearings of the reduction gear and nose section. Additionally, as seen in Figure 4.9, oil passages (120) delivered oil under pressure to the valve gear and camshaft gear train (60-63) bearings associated with each cylinder bank.

Within each camshaft gear (63), as shown in Figure 4.12, an automatic camshaft oil pressure reducer admitted oil to the hollow camshaft (121) at a pressure less than full pump pressure. The purpose of this reduction in oil pressure to the camshaft was to control oil flow and to avoid excessive oil bleed from the pressure system, since the quantity of oil required by the valve gear was small.

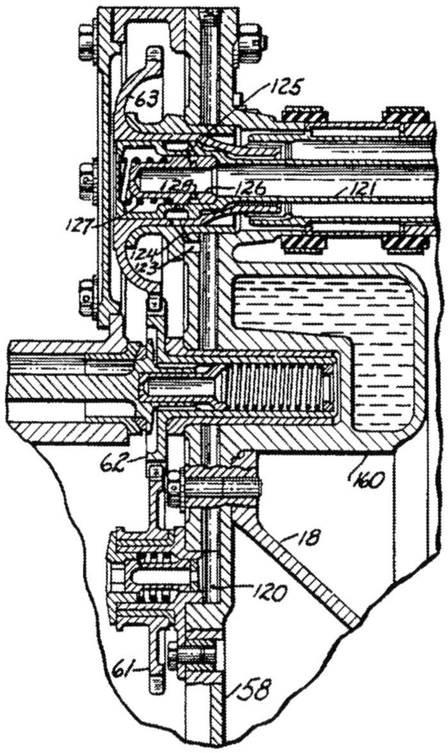

Figure 4.12. Enlarged Longitudinal Section through the Engine Front Section showing the Camshaft Oil Pressure Reducer (from U.S. Patent No. 2,355,277)

Key to Figure 4.12

18 – Front crankcase
58 – Idler gear support diaphragm
61 – Idler gear
62 – Idler gear
63 – Camshaft gear
118 – Drilled oil passages
120 – Oil passage
121 – Camshaft
123 – Camshaft oil passage
124 – Camshaft oil passage
125 – Camshaft oil passage
126 – Camshaft oil pressure reducer valve
127 – Camshaft oil pressure reducer valve spring
128 – Reducer valve oil passage
160 – Front annular coolant manifold

To this end, drillings (123, 124, and 125) were made in the bushings and hub elements of the camshaft gear. A sliding valve (126) was normally forced to the right by a spring. Within this valve were radial drillings (128), which at times registered with drilling (125) to allow entrance of oil into the hollow camshaft (121). When the oil pressure within the camshaft increased, the valve (126) was forced to the left against the spring (127) and throttled the drilling (125). This action maintained within the camshaft an oil pressure controlled by the strength of the spring (127).

In the schematic representation of the lubrication system in Figure 4.11, this oil pressure reducer was designated (126). Another pressure reducing valve (129), shown in Figure 4.11, reduced pressure between the rear oil pump (115) and a low-pressure oil annulus (130) in the accessory drive section. This annulus served as an oil header for the lubrication of the accessory driven gears.

Scavenge Circuit

Scavenge oil flowed by gravity to the lower portions of the crankcase and to the lower valve cavities. Each valve cavity was equipped with two fittings located at the top and bottom front corners of the cavity. One fitting was also provided at the rear lower corner of each valve cavity.

Each front fitting was connected via a hose to the fitting of the adjacent cylinder across the vee formed between the banks. In this fashion, scavenge oil drained from the top down and ultimately wound up in the valve cavities of the two lowest cylinder banks, "D" and "E." From here it was combined with scavenge oil from the front part of the crankcase and nose case. It then flowed through scavenge duct (Figure 4.1 — 12) to the scavenge portion of the front oil pump (96) and was pumped from the engine. Scavenge oil from the rear of each valve cavity drained via a hose attached between the fitting in its rear lower corner and the rear accessory section. Scavenge oil from the valve cavities, rear crankcase, and rear accessory section was combined and pumped from the engine by the scavenge portion of the rear oil pump (115).

Cooling System

The Tornado liquid cooling system is depicted schematically in Figure 4.13 and 4.14. Elements also appear in Figures 4.1 through 4.6. Two centrifugal coolant pumps (154) were located on either side of the aft lower face of the accessory section. Coolant liquid, a mix of approximately equal parts of water and ethylene glycol, was placed under pressure by the coolant pumps (154) which were connected by coolant pump discharge conduits (155) to a rear annular coolant manifold (156) cored in the rear supercharger section of the engine casing. From this manifold, coolant flowed through rear coolant conduits (157) to the upper part of the cylinder bank cooling jacket of each cylinder bank. The coolant flowed longitudinally forward through the cylinder bank cooling jackets, through front coolant conduits (158), and was collected in the front annular coolant manifold (160). This manifold was cored in the idler gear support diaphragm (58) casting.

Coolant from the front annular coolant manifold (160) was returned through a pipe (162) to a radiator (163) and finally back to the coolant pump intakes through pipes (164). To allow for coolant expansion and for complete filling of the whole coolant system, a header tank (166) was mounted above the engine and connected by pipes (167 and 168) to the uppermost portions of the front and rear annular coolant manifolds (160 and 156).

Since coolant was delivered to and removed from the uppermost parts of the engine, provisions were included to avoid vapor locks in the system. Figures 4.6, 4.13 and 4.14 show cylinder bank coolant connectors (170) that were placed between the jackets of adjacent cylinder banks at the most forward and most rearward locations of each cylinder band. This allowed air, gas, or steam to flow upward through successive cylinder jackets to the uppermost jacket and to be removed via the front and rear annular coolant manifolds (160 and 156) to the header tank.

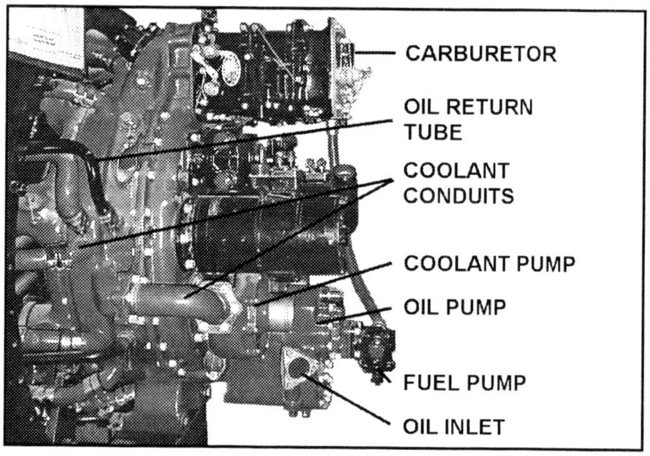

Rear of Engine Showing Accessories, Coolant, and Lubrication Components

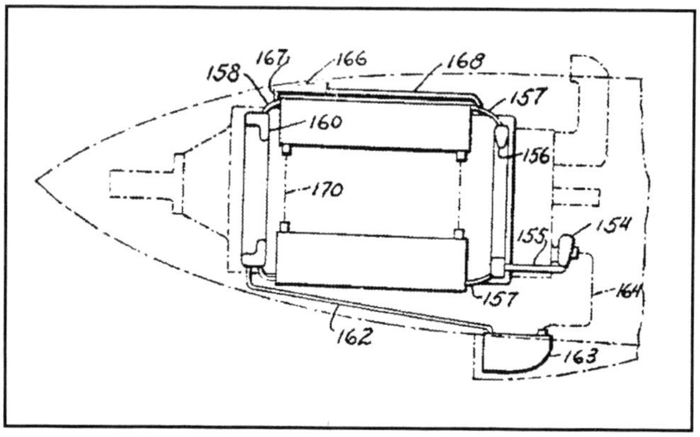

Figure 4.13. Side View of the Engine Installation Showing Cooling System Details (from U.S. Patent No. 2,355,277)

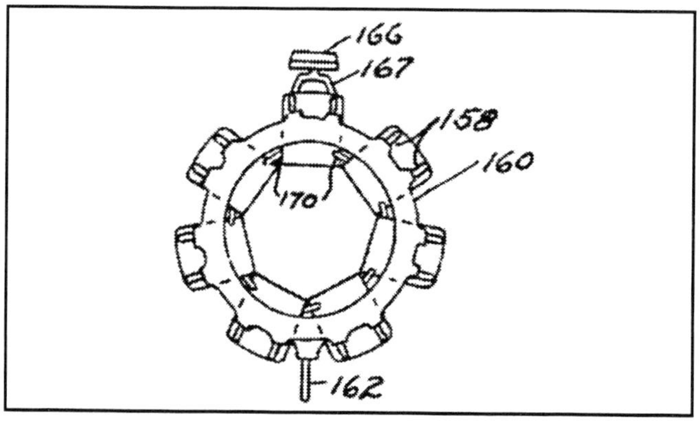

Figure 4.14. Front View of the Engine Showing Cooling System Components (from U.S. Patent No. 2,355,277)

Key to Figures 4.13 and 4.14

154 – Coolant pumps
155 – Pump discharge conduits
156 – Rear annular coolant manifold
157 – Rear coolant conduits
158 – Front coolant conduits
160 – Front annular coolant manifold
162 – Coolant return pipe
163 – Radiator
164 – Coolant pump inlet pipes
166 – Header tank
167 – Header tank pipe
168 – Header tank pipe
170 – Cylinder bank coolant connectors

Left Side View of Engine Showing Accessories (Courtesy of the New Jersey Aviation Hall of Fame and Museum)

Top View of Engine Showing Accessories and Plumbing Connections (National Air and Space Museum, Smithsonian Institution, SI Neg. No. 2001-831)

Bottom View of Engine Showing Plumbing Connections (National Air and Space Museum, Smithsonian Institution, SI Neg. No. 94-13743)

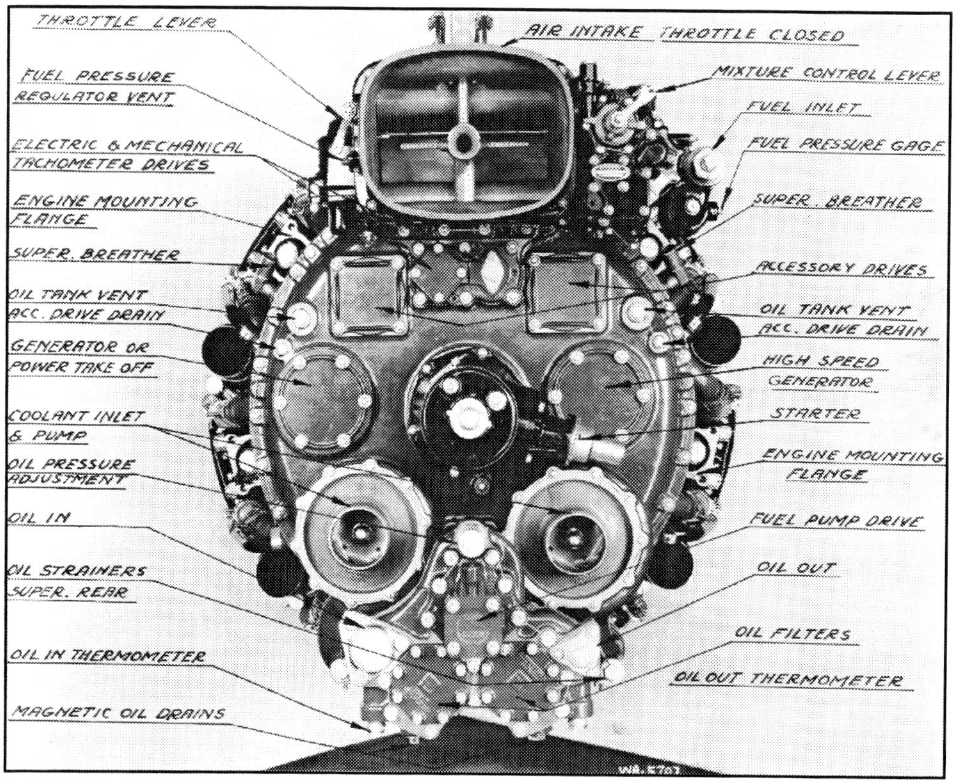

Rear View of Engine Showing Accessories and Plumbing Connections (National Air and Space Museum, Smithsonian Institution SI Neg. No. 2001-828)

Ignition System and Fuel Injection System

As shown in Figure 4.1, the T-42 was equipped with seven Bendix-Scintilla Model DF6L magnetos (65), located at the front of each cylinder bank, which ran at 0.75 times crankshaft speed. Each dual magneto was flange mounted and operated with a left-hand rotation when viewed from the flange end. Each magneto fired all 12 spark plugs in the six-cylinder bank to which it was attached.

A dual magneto is unique in that two coils, two condensers, two distributors, and two sets of points are arranged around a single rotating magnet and drive train. The dual design duplicated the relatively unreliable electrical components while retaining only one of the more expensive and reliable rotating shaft, bearing set, rotating magnet, and drive components. The DF6L was also unique in that the shaft driving the two distributors is perpendicular to the rotating magnet centerline. This allowed for a very compact magneto enclosure and facilitates further speed reduction of the distributor so that 1.5 revolutions of the rotating magnet were required for each revolution of the distributor. The DF6L magneto was built specifically for the Tornado. Since only six complete engines were built, it is unlikely that more than 50 of the magnetos were used.

The magneto harnesses were also custom items. They included the distributor cap and an exterior shield that supplies cooling air to the hot and cramped space below the exhaust manifolds. The only magnetos and harnesses known to exist today were on the engine displayed at the New Jersey Aviation Hall of Fame and Museum in Teterboro.

If the engine was a fuel-injected model, a fuel injector unit (68) was attached to each cylinder bank and driven from each camshaft gear (63). The magneto and fuel injector of each cylinder bank served only that bank and could be timed in perfect harmony with the piston positions of that bank. This arrangement avoided the timing inaccuracies common to most radial engines with crankshaft driven magnetos serving all cylinders. In the Tornado, one magneto serves all master rod cylinders, and another magneto serves each aligned bank of articulated rod cylinders.

The Tornado firing order changed at least once during its development. Entries in a field service engineer's notebook indicate the first engine fired the cylinders in each bank sequentially (1-2-3-4-5-6). This is corroborated by the firing order discussion in the Stress Analysis Report. Later engines fired the cylinders in each bank in a staggered order (1-2-5-6-3-4). It is not known why these changes were made. More information on the complete Tornado firing order is included in Appendix E—"Data from Engine No. 11770." A schematic of the drive train for the

The DF6L magneto with one coil removed showing internal configuration. (Courtesy of the New Jersey Aviation Hall of Fame and Museum)

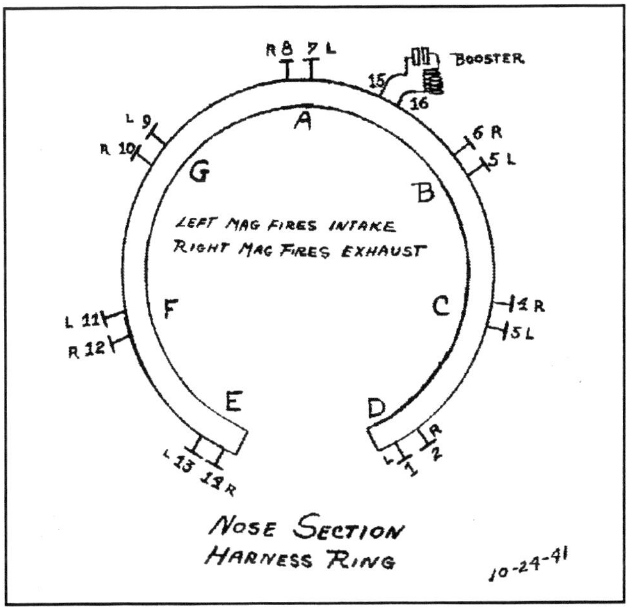

A sketch of the nose section ignition wiring harness ring (looking aft) shows magneto primary lead connections. (Courtesy of the New Jersey Aviation Hall of Fame and Museum)

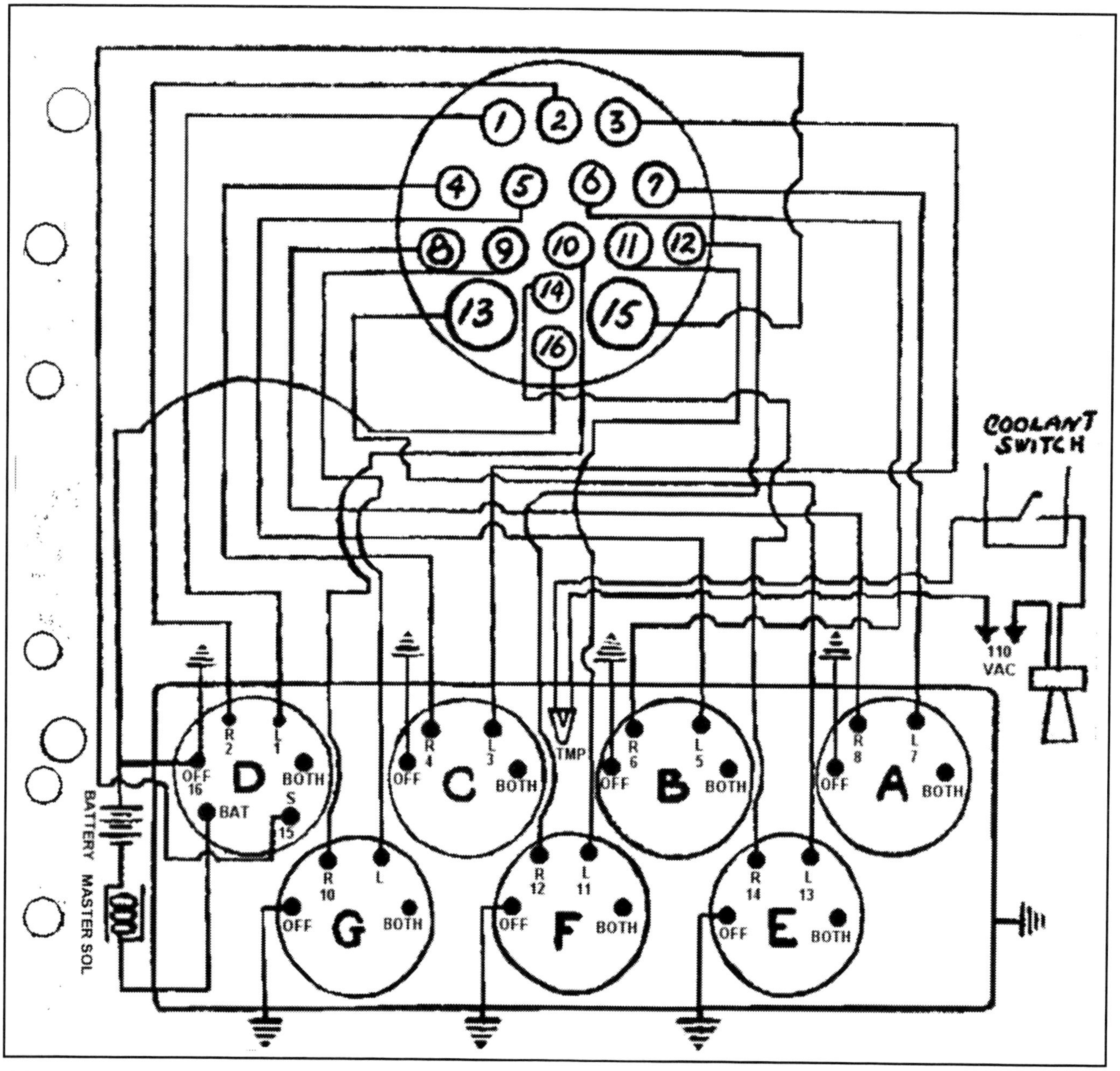

This pencil sketch is from the notebook of Wright Aero Field Engineer John McElroy. Shown is the ignition system schematic for engines tested at Paterson, New Jersey. (Courtesy of the New Jersey Aviation Hall of Fame and Museum)

camshaft, fuel injector unit, and magneto is shown in Figure 4.10.

As can be seen from the forgoing discussion, the Tornado designers met their objectives of a compact and lightweight engine by innovative engineering, clever design, and elegant packaging. Any new engine has its growing pains, and the Tornado was no exception. However, for an engine so revolutionary and complex, technical development of the Tornado actually proceeded reasonably well. As will be seen in the next chapter, the same could not be said of non-technical issues surrounding the engine.

---

1 Unless otherwise noted, all material related to T-42 design and construction comes from study of: U.S. Patents by Daub and Goodman; Goodman, *Flat and Radial Design Studies*; Hunter, *Supplement No. 1*; Lundquist, *Tornado Design Studies*, and *Stress Analysis of R-2160*.

View of Engine Front Showing Magnetos, Low and High-tension Harnesses, Harness Cooling Shrouds, and Fuel Injection Pump Drive Pads

Left: This details the busy area forward of the cylinder banks. Silver ignition shrouds are clearly seen top and bottom. In the center, with clamps on either side, is a passage containing the camshaft drive coupling. Directly above and below the camshaft are coolant passages. Above and below coolant passages are black oil drain lines from the camshaft chamber. Just in front of the large black cylinder cover (left) are the front cylinder hold-down studs and nuts. One marvels at the large number of clamps and hoses, and wonders just how reliable such an arrangement would have been.

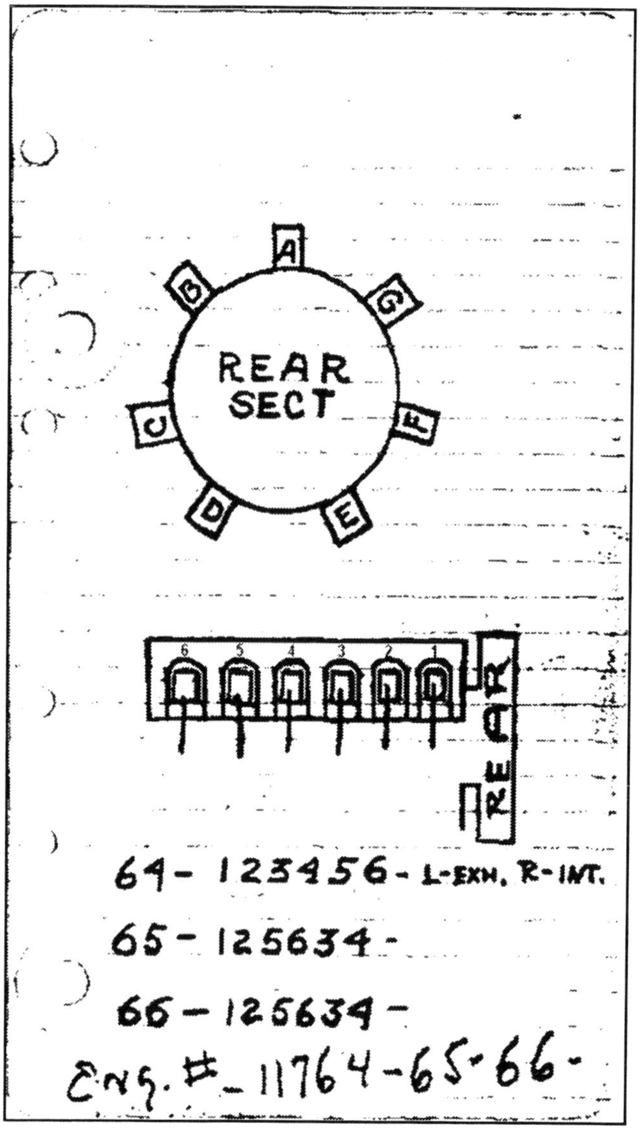

Another of the pages from McElroy's notebook shows cylinder numbering. Note the revised firing order for engines 11765 and 11766 (Courtesy of the New Jersey Aviation Hall of Fame and Museum)

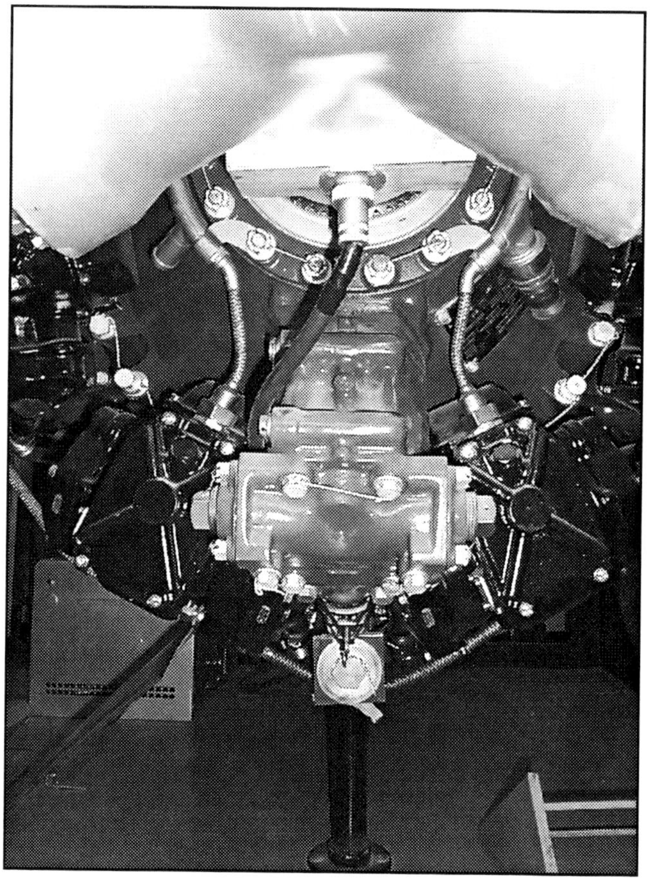

Clearly visible in this photograph is the front oil pump. This pump served the dual role of supplying pressurized oil to the forward portions of the engine plus scavenging the nose case, camshaft chambers, and drive trains.

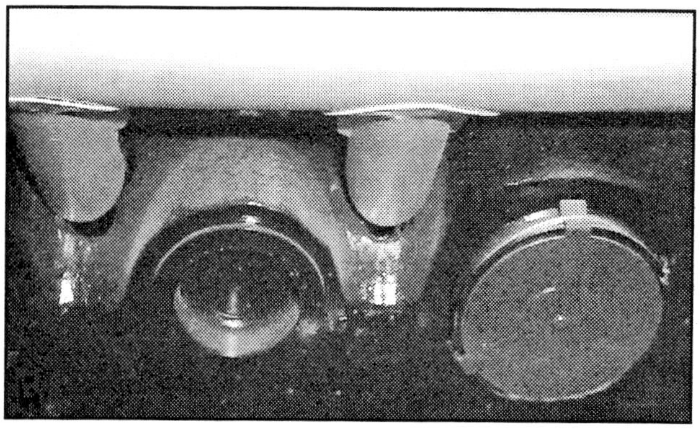

Left: Detailed here are two spark plugs surrounded by an ignition cooling shroud. The cover plate on the lower right of the photograph blocks an exhaust port. The circular area in the lower left of the image is a water jacket plug. Because of their close proximity to the exhaust ducts, ignition leads had to be protected by the cooling shrouds. In addition to reflecting radiant heat, the cooling shrouds ducted cool air to the spark plugs and ignition harness.

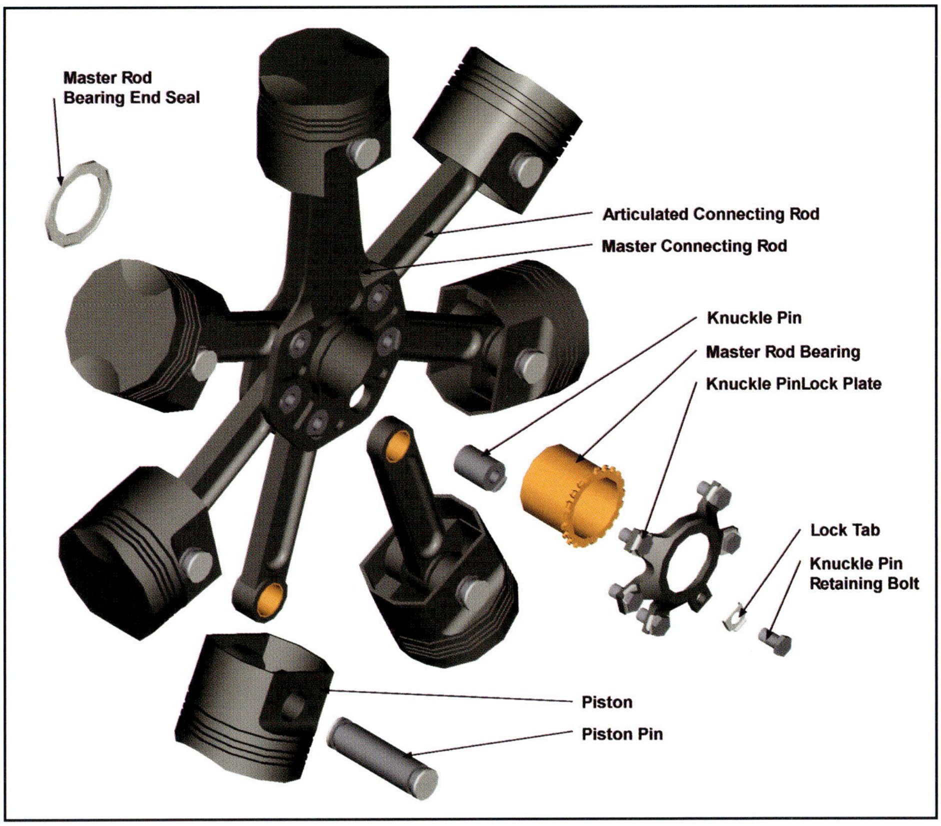

Except for shorter rod lengths, the Tornado connecting rods and pistons followed standard Wright Aero practice.

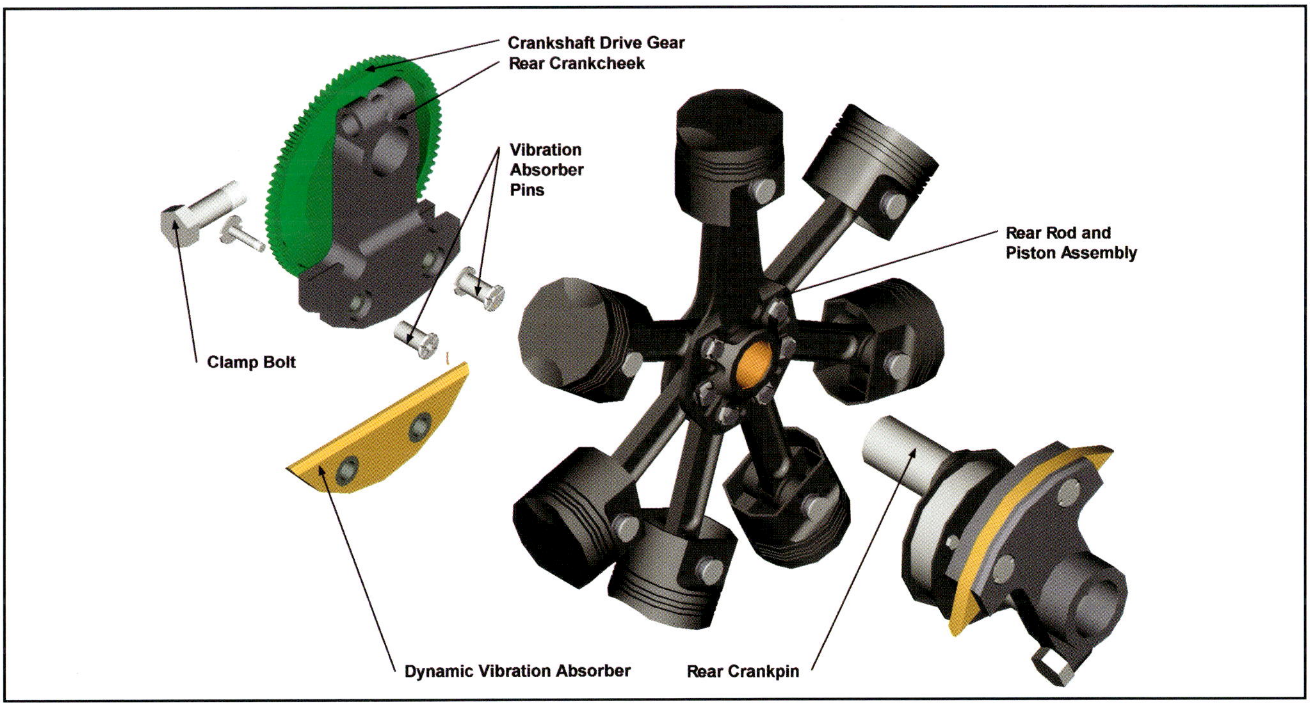

As was typical of other Wright engines, one-piece master rods were assembled onto the crankpins of built-up crankshafts. Friction generated by enormous clamping forces held the crankshaft together. Dynamic vibration absorbers damped second-order torsional vibration.

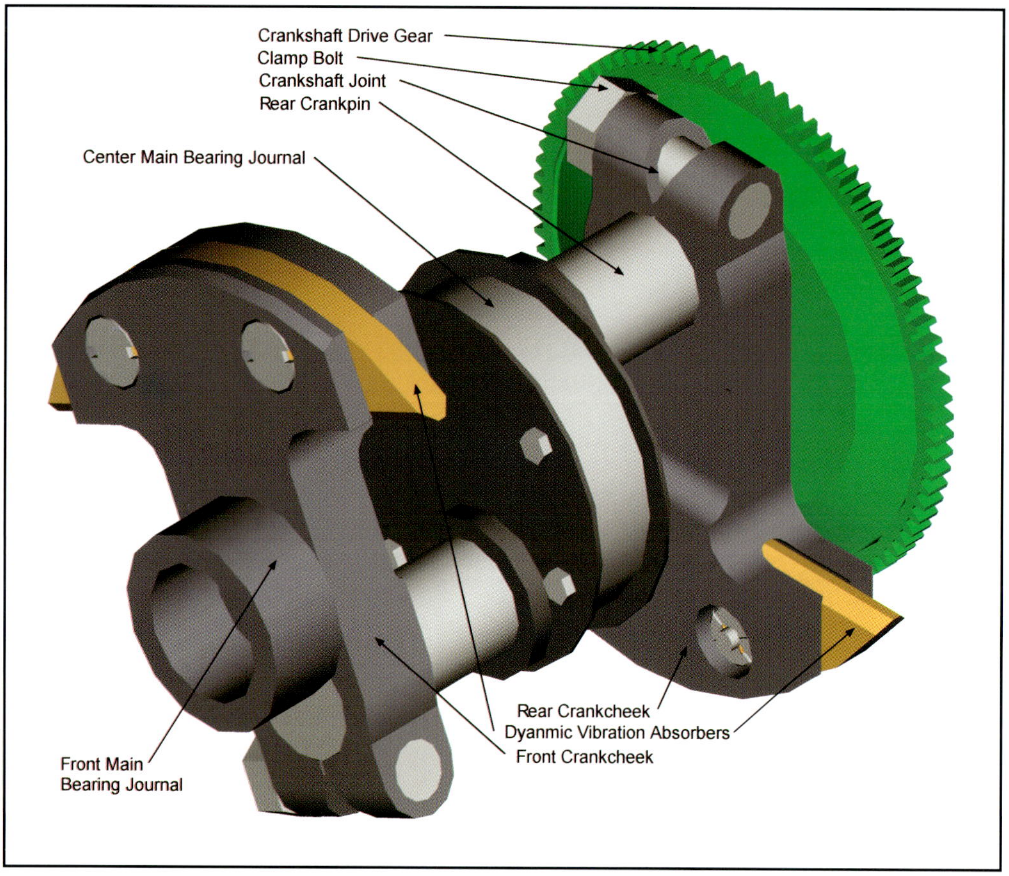

Small increases in crankpin diameter result in large increases in engine diameter. By sizing the crankpins to handle only the power produced by a two-row group of 14 cylinders, Tornado designers were able to hold engine diameter to 36.5 inches. Power from each group of cylinders was transferred to the propeller and supercharger via the large gear at the rear of each crankshaft unit. Crankpins were 154 2/7 degrees apart, producing even firing intervals for each group of 14 cylinders.

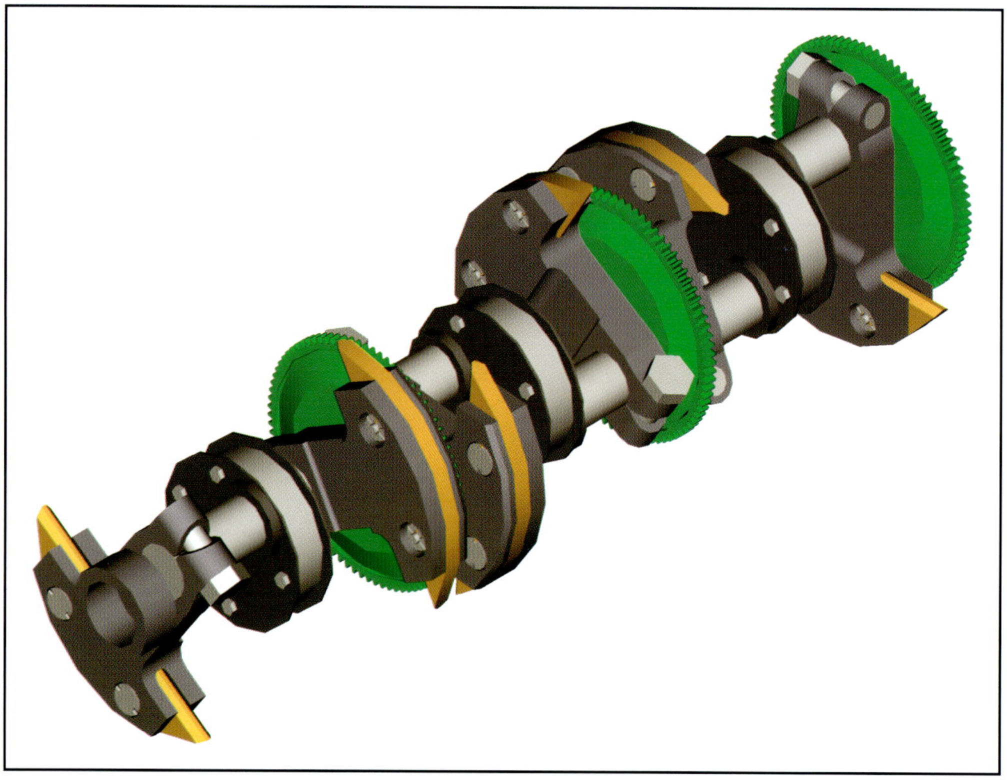

Three crankshaft units were placed in tandem and phased at 120 degrees to one another, producing even firing intervals for all 42 cylinders. Although the crankshaft units shared front and rear main bearing journals, they were free to twist slightly in relation to one another. This effectively isolated each group of 14 cylinders and greatly reduced torsional vibration.

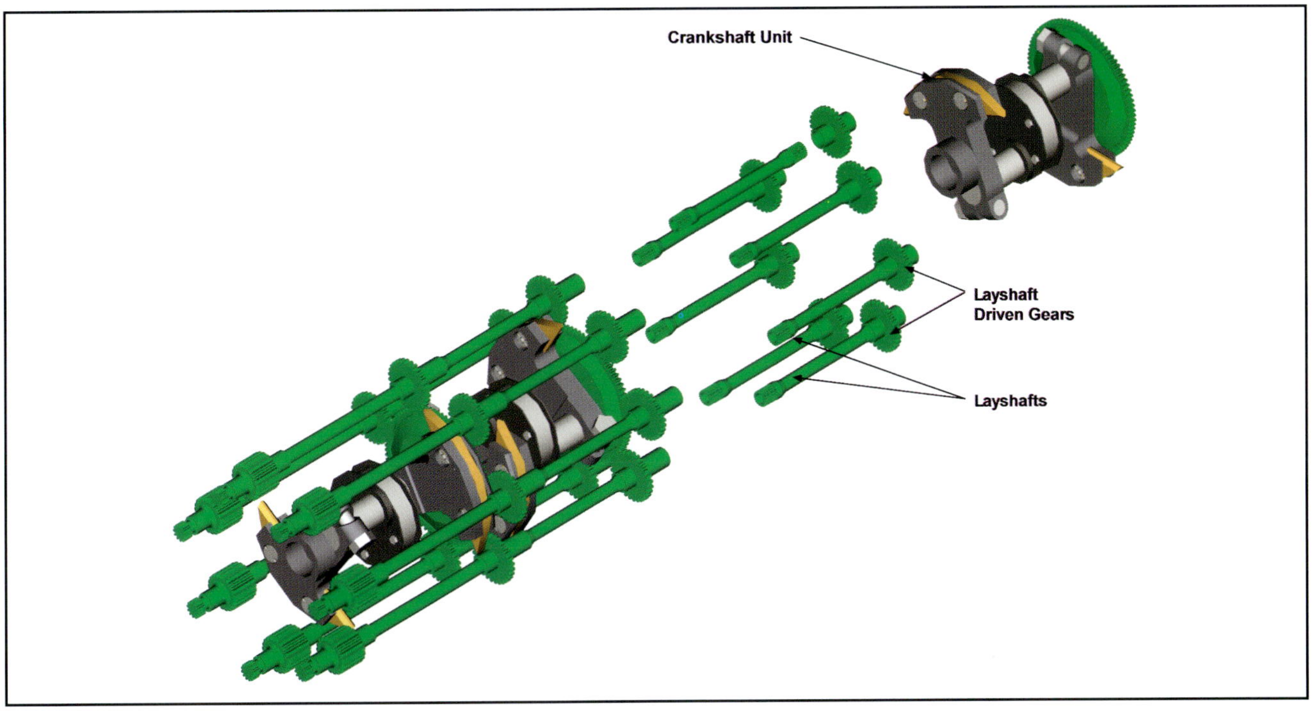

Seven layshafts transfer power from the crankshaft units to the propeller reduction gear and supercharger. The layshafts are torsionally resilient, which helps isolate vibration and ensure that all gear teeth are equally loaded.

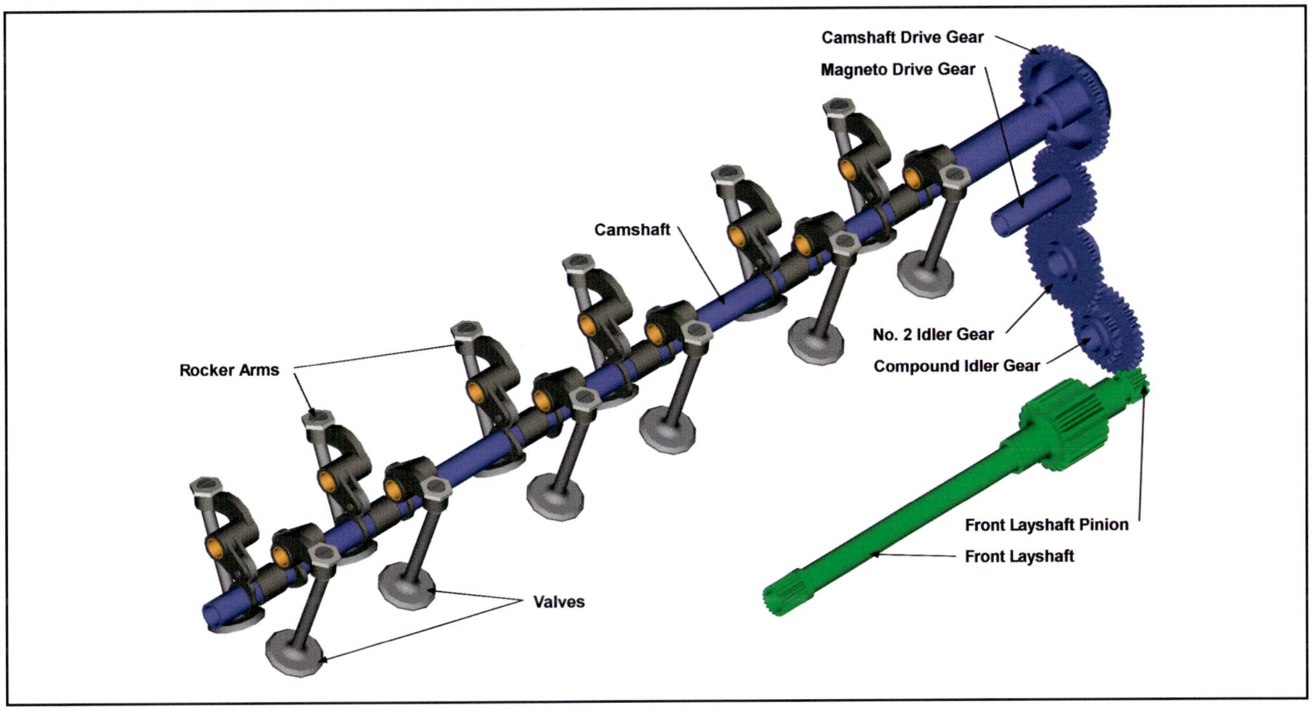

Each of the seven overhead camshafts is driven via a gear train originating at a front layshaft segment. Each gear train also drives a magneto. Rocker arms have rollers that run on the cam lobes and a screw for adjusting valve clearances.

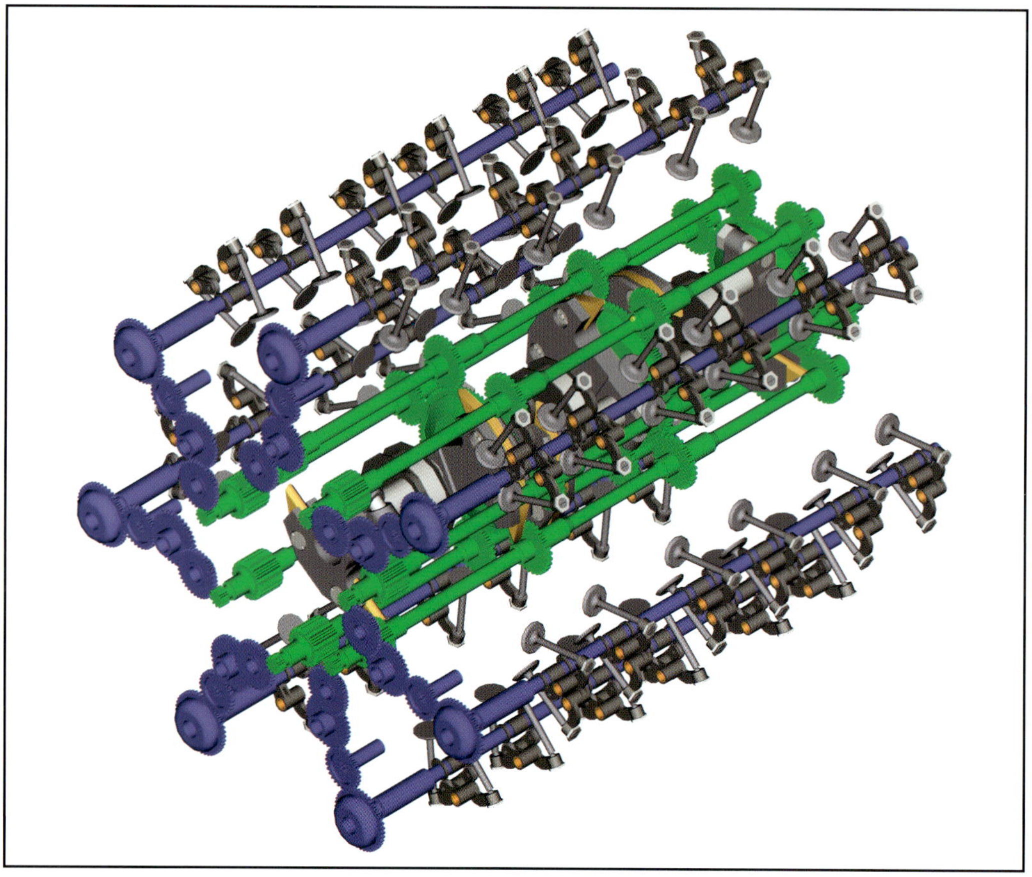

Crankshafts, layshafts, and valve gear comprise the rotating components of the power section. In this view, connecting rods and pistons are omitted for clarity.

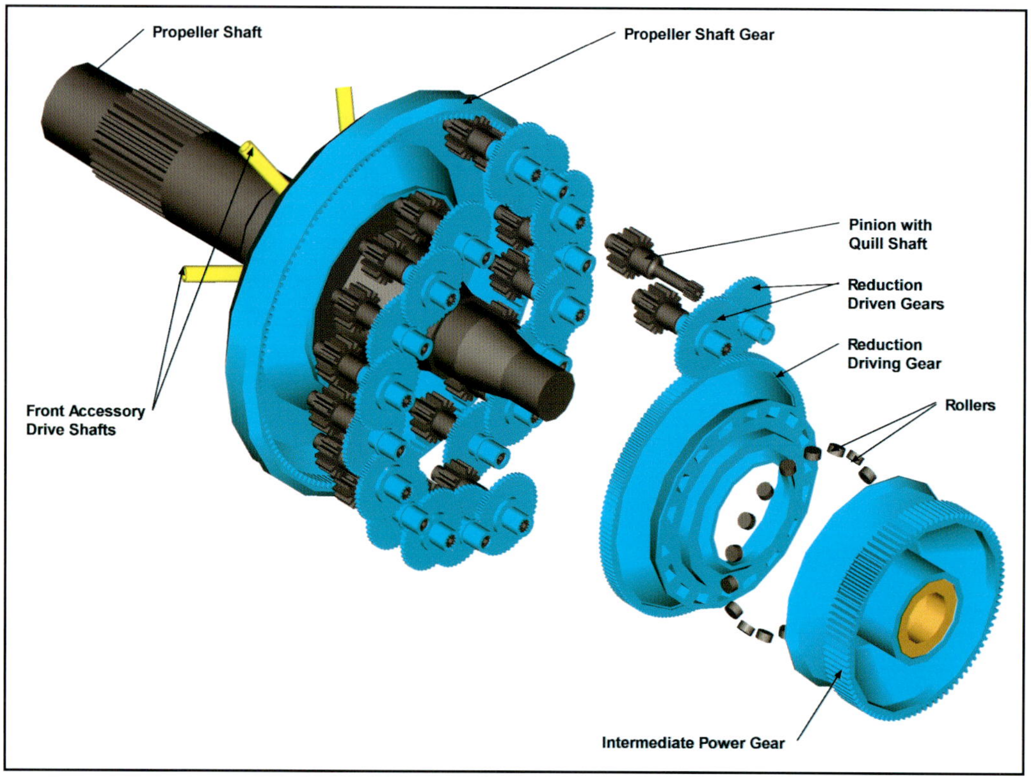

The propeller reduction gear with integral torquemeter was an elegant solution to the problem of transferring up to 2,500 hp to the propeller. Provisions existed to drive the front oil pump, a propeller governor, a gun synchronizer and two other front accessories.

The Tornado supercharger consumed more than 200 hp, making its drive design a non-trival endeavor. Torsionally flexible quill shafts connected each of the seven supercharger driving gears to a layshaft. Like the layshafts themselves, and the quill shafts in the propeller reduction gear, these quill shafts isolated vibrations and ensured equal gear tooth loading.

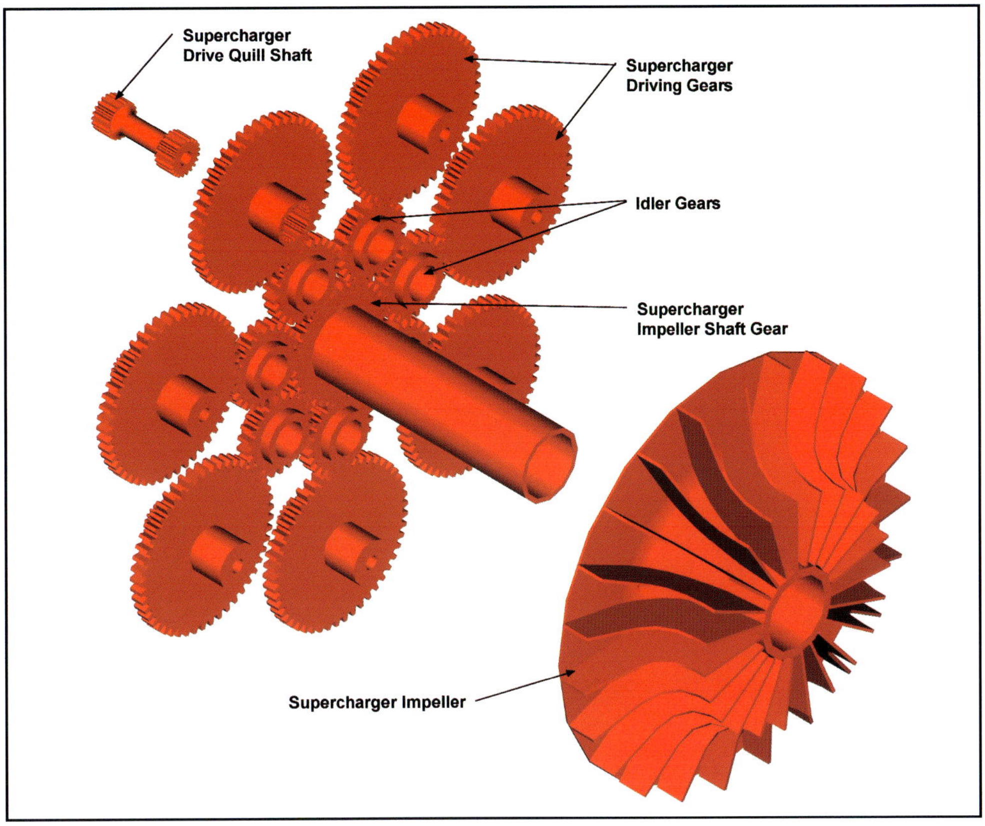

Power for all rear accessories, along with a means of connecting the electric starter, was provided through a crankshaft extension shaft. A gear train provided an appropriate drive ratio for each accessory.

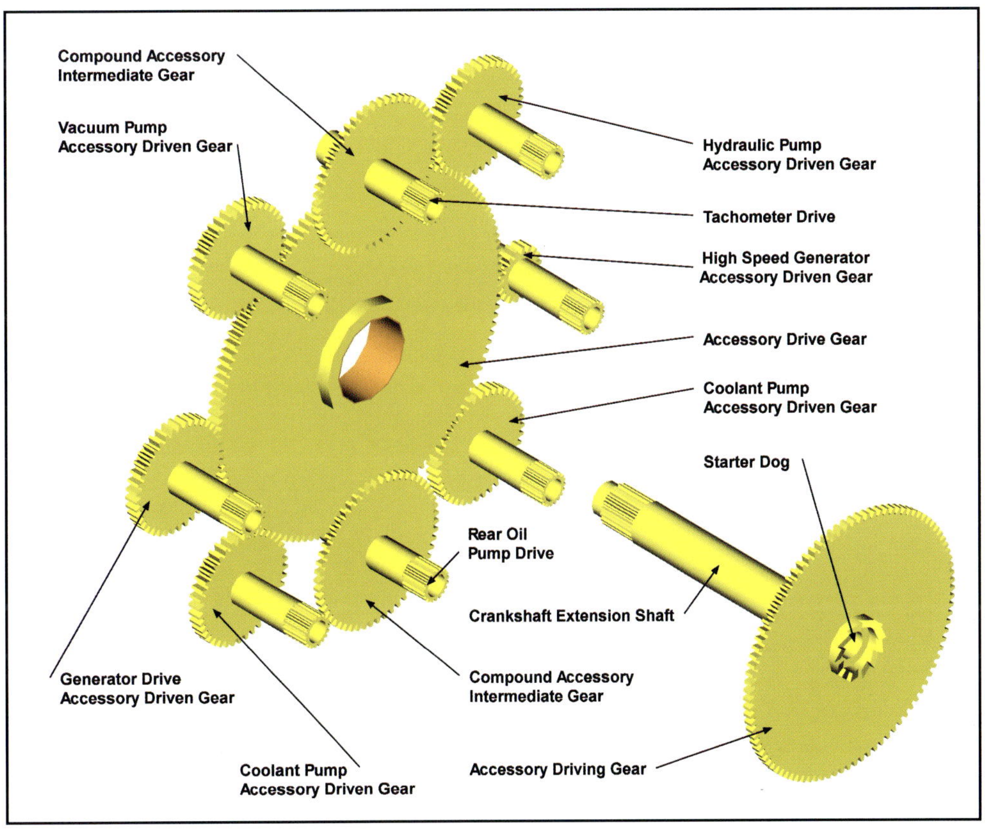

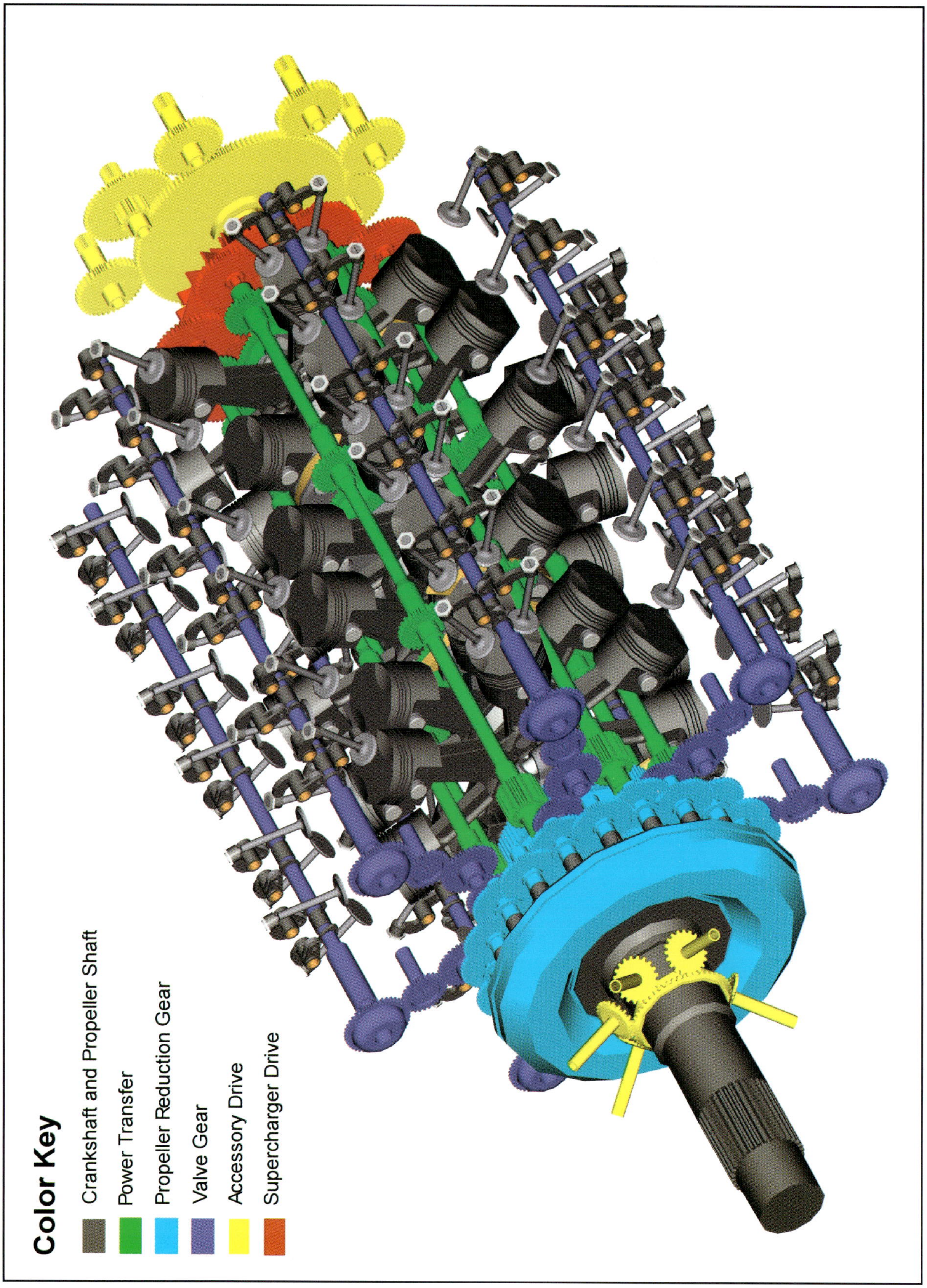

This left front view of the complete engine shows the integration of all major components.

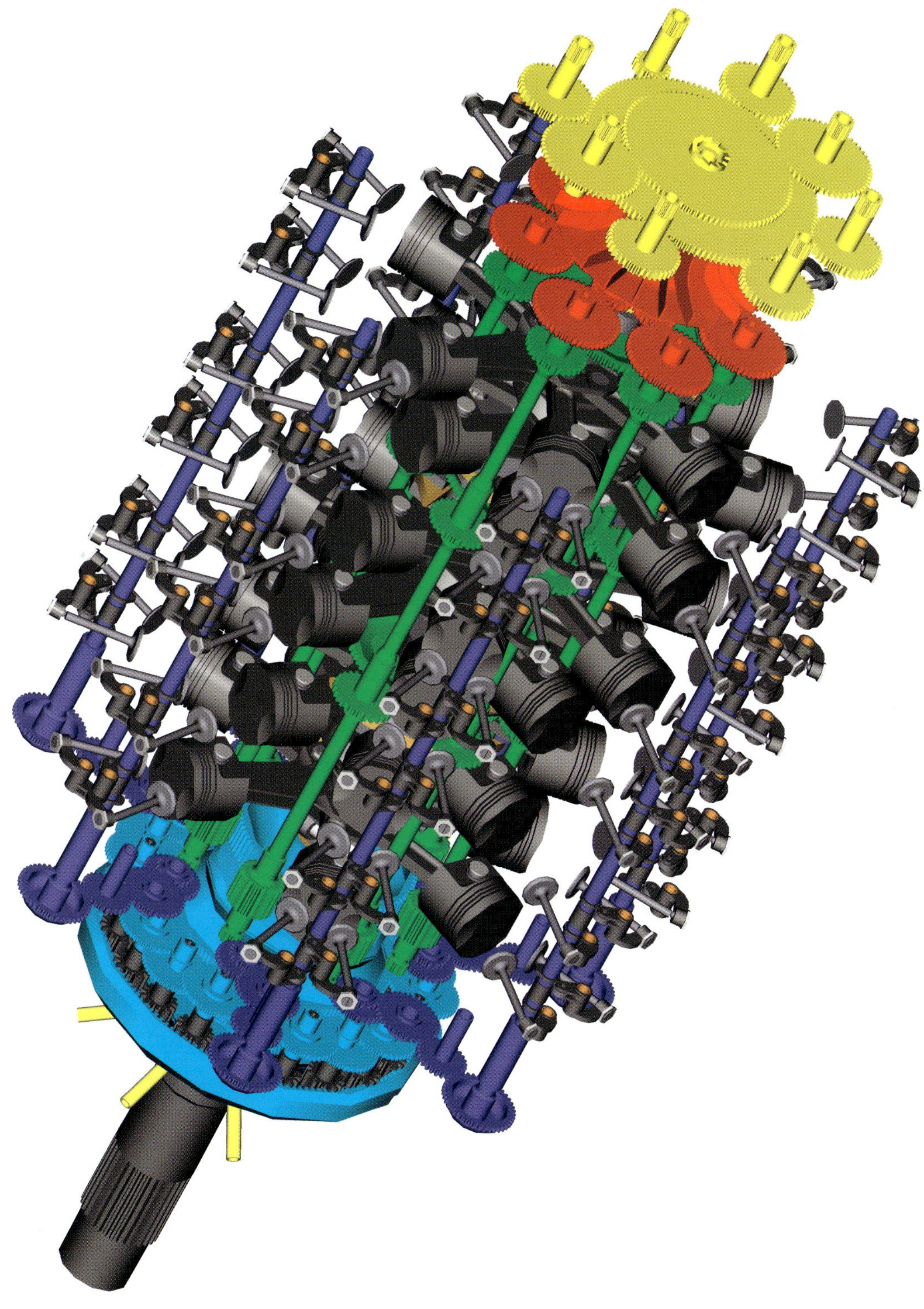

Rear View of the Complete Engine

Front Accessories, Wiring, and Plumbing

Rear Accessories and Plumbing

The last remaining Tornado 42 is displayed at the New Jersey Aviation Hall of Fame and Museum.

5 The Tornado Program

A Rear View of the Tornado 42. This was one of several "publicity" photographs circulated by Wright Aero. (Arthur Nutt Papers, Acc. 1987-0015, National Air and Space Museum, Smithsonian Institution, SI Neg. No. 2001-2658)

The contractual, political, and technical issues surrounding the Tornado development provide interesting insight into the nature of the Army and Wright Aero around the time World War II began. It is also practically the only account of technical challenges faced by the design team. The following is a chronological account of the Tornado program.

The Tornado program began officially on April 12, 1939, when the Army Power Plant Lab at Wright Field, Ohio, initiated Authority for Purchase (AFP) No. 148115. This resulted in contract W535 AC-12632, which, after some revision, was approved on June 24, 1939. The contract specified that Wright Aero was to build a 14-cylinder portion of the larger 42-cylinder engine and deliver two reports on 50-hour and 150-hour endurance tests of the Tornado 14 (T-14). Cost was $72,900.[1]

By late May 1939, Wright Aero had decided to proceed with the Tornado 42 (T-42) in parallel with the T-14. The T-42 was to be completed about mid-September 1940 – three months ahead of the predicted completion of T-14 testing.[2] On January 23, 1940, contract W535 AC-13709 for one R-2160-1 was approved at a cost of $278,000.[3]

During the first half of 1940, Wright Aero was busy with engineering activities on both the T-14 and T-42. On February 13, 1940, the first report on Tornado design studies was released.[4] This brief eight-page report, a deliverable of contract W 535 AC-13709, served primarily as a cover for two assembly drawings. Drawing No.413460 was a longitudinal cross section of the engine. Drawing No.LS-5302 was a transverse section. A stress analysis was provided under separate cover.

It was reported that no single cylinder testing had been done or was contemplated. Parts for four T-14 engines were being fabricated, and all testing would be done with these. The first of the T-14s had been completed and was being motored over at speeds as high as 4,900 rpm. It was anticipated that this T-14 engine would be outfitted for powered operation by late February 1940, and would then proceed directly to endurance testing.

Pick a Reduction Gear, Any Reduction Gear

Beginning in August of 1939, there was substantial Army interest in a variety of propeller reduction gears. The Materiel Command in Washington had decreed that all large engines were to be designed with a variety of reduction gear options. First, they wanted two-speed reduction gears that allowed different propeller gear ratios for takeoff and cruising. Second, they needed support for contra-rotating propellers that eliminated torque and propeller diameter concerns when large engines were installed in small pursuit aircraft. Finally, they needed left and right rotation for multi-engine applications that did not utilize contra-rotating propellers. Further complicating the myriad gear options was the need to provide different gear ratios to suit varying aircraft speeds, and the need for the propeller shaft to accommodate both Curtiss electric propellers and Hamilton Standard hydraulic propellers. Additional items on the propeller reduction gear wish list were provisions for a torquemeter and lubrication of the nose case separate from the engine oil supply. Wright Aero put a substantial amount of effort in specifications, preliminary designs, and cost exercises for gearboxes that could be used across the board in the Wright R-2160, R-2600, and R-3350 engines.

When Wright asked for an additional $150,000 for this work, the Army decided to wait and see if the Tornado materialized before pursuing the two-speed, two propeller options. Single speed, left and right rotation was pursued for the Tornado.[57] This is one of many examples of Army requirements that sounded nice at the time but never panned out in practice. The author is aware of no actual production application of two-speed gearboxes and no successful production use of contra-rotating propellers on reciprocating engines built in the United States. Like the provisions for a gun synchronizer that was required for every U.S. aircraft engine, complex gearboxes were a classic example of military over-specification.

AFP No. 168260, initiated on August 21, 1940, authorized Wright Aero to develop two-speed dual-rotation reduction gears and to conduct 150-hour endurance tests on the unit. These reduction units were to be used both on the T-42, the R-2600, and the R-3350. Contract W535 AC-15738 resulted.[58]

The first installment of the stress analysis was delivered along with the *Tornado Design Studies*.[5] The object of this activity was to analyze the crankshaft and valve gear and to determine the bearing loads and torsional vibration characteristics. Using Air Service Information Circular No. 421 as a guide, inertial and gas forces were calculated or determined using graphical methods. This exercise did not identify any specific problems with the engine.

The second installment of the stress analysis was issued during July 1940.[6] The connecting rods, piston and knuckle pins, power shafts, propeller reduction gears, and camshaft drive train were analyzed by calculation and graphical methods. As before, no specific weaknesses were identified. Conspicuously absent was any stress analysis of either the accessory or supercharger drive train. It is not known whether this was not done, not reported, or whether the reports were lost.

On June 20, 1940, Army decided that since General Electric did not yet have a turbosupercharger available, altitude testing of the engine would wait until the turbo and new altitude test laboratory were complete.[7]

Progress Report—The Wright Aero Conference, July 31 and August 1, 1940

On August 1, 1940, Wright Aero reported that the two T-14s had accumulated 85 and 112 hours, respectively. The first unit, Serial No. 11760, had been used for oil scavenging, take-off, and endurance runs and had run 22.5 hours at 400 hp, 24.7 at 500 hp, 3.8 hours at 600 hp, and 0.3 hour at 700 hp. The second unit, Serial No. 11761, had completed the take-off test and was used to develop indicator card data and for general run-in of parts. It had accumulated 67.2 hours at 400 hp, 7.4 hours at 500 hp, 9.7 hours at 600 hp, 1.2 hours at 700 hp, and 26.2 hours motoring over. The only significant trouble was with main bearings due to excessive deflection in the main bearing flanges. Redesigned parts were fabricated and solved the problem.[8]

This progress report came at a conference held with Tornado project engineer W. G. Lundquist of Wright Aero at its Paterson, New Jersey, plant. Wright Aero promised information on weights and drawings of the T-42 engine mount in about seven days time. The first cylinder banks were cast of bronze by an outside vendor. Work on aluminum cylinder banks would follow. It was planned that the first T-42 would start running on September 1, 1940. Delivery of the first engine to Wright Field was to be on October 15, 1940.

Army personnel viewed a complete set of wooden mockups of the T-42. Outstanding features included extreme compactness and the mounting of seven magnetos on the front face of the seven cylinder banks. Each magneto supplied dual ignition to the six cylinders of each bank. Induction pipes of elliptical cross section were neatly arranged in the vee between cylinder banks. Each induction pipe was split along the major axis, with each half feeding three cylinders. Exhaust stacks were arranged above the intakes. Fabrication of the spark plug leads remained to be done.

There was some discussion about whether the T-42 would have "true motion" rods. Lundquist stated that true motion rods would not be used. Mr. R. W. Young, Wright Aero vice president of engineering, recounted work Wright had done with true motion rods for the Navy on the X-1670 and stated that since Wright had numerous priority projects, no effort would be made at that time to adapt true motion rods.[9]

On September 18, 1940, at the request of Gen. Ira C. Eaker, then-Col. Oliver P. Echols recommended authorization for Howard Hughes to buy two T-42s for the DX-2, an all-wood twin-engine pursuit/attack aircraft privately funded by Hughes. It was impossible to predict delivery dates for these engines.[10]

The Successful 50-Hour Development Test of the T-14

A 50-hour development test of the T-14 was conducted at Wright Aero from July 23 to September 23, 1940. As might have been expected of an engine with so little running time, there were problems, but no major ones.

W. G. Lundquist reported on September 19, 1940, that T-14 No. 11760 had been torn down after a

A Complexity Comparison

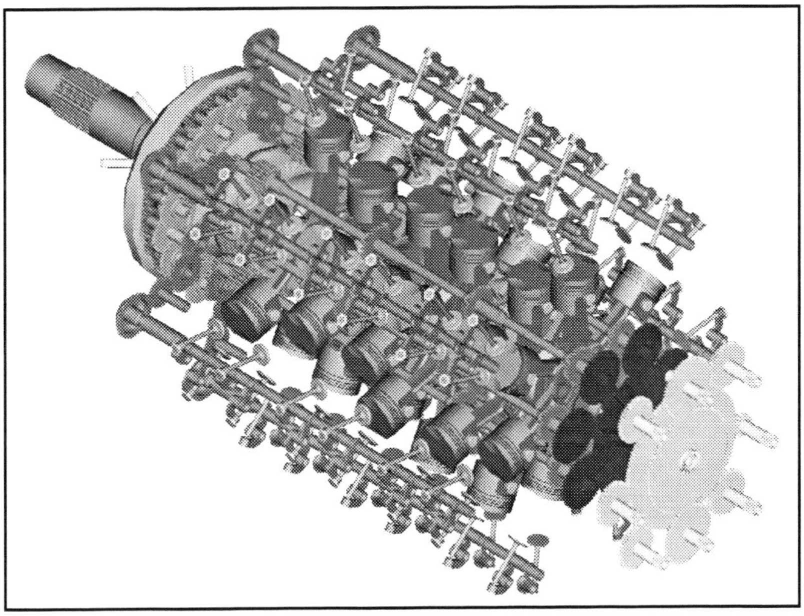

The psychological impact of 42 tiny cylinders and myriad other parts that made up the Tornado was a liability. On May 12, 1941, Wright Aero sent the Army a comparison of part counts and production costs for the Tornado, the R-2600, and the R-3350. Details are presented in Table 5.1. Table 5.2 is a manufacturing cost comparison. The low manufacturing cost per horsepower ($5.77) for the Tornado was due to the low number of different detailed parts. Wright felt that, despite the large number of cylinders, the relatively smaller number of different detailed parts could be manufactured cheaply in volume production. In addition, since no large or heavy parts were included, smaller production machinery could be used. The Tornado was about the same installation size as Allison and Rolls Royce engines, but had approximately two and one-half times the horsepower.[59]

Table 5-1. Comparison of Parts Countsfor Several Wright Aero Engines

Number of	R-2160	R-2600	R-3350
Different detail parts	725	720	810
Different assemblies	150	130	105
Different part numbers	875	850	915
Total parts per engine	10,200	8,000	10,150
Standard parts	5,765	4,820	6,425

Table 5-2. Comparison of Manufacturing and Sales Costsfor Several Wright Aero Engines

Model	Mfg. Cost per HPr	Sales Cost per HP
R-2600-A	$5.78	$9.85
R-2600-B	$6.29	$9.55
R-1820 (G200)	$5.87	$8.58
R-2160	$5.77 (estimated)	
R-3350	$6.10	$9.77

short run to determine the cause of an oil pressure drop. Two oil plugs were left out of the crankshaft, preventing oil pressure from building. The plugs were replaced, the engine reassembled, and running started on September 20, 1940. T-14 No. 11761 had been reassembled and was again running, making special checks on valve cams, changing the overlap of the valves from 105° to 30°. This change had resulted in a loss of about 50 hp, necessitating additional study on cams. New cams with a 1/8-inch increase in valve lift were being procured. This engine had been operated at 630 hp, 42 inches of manifold pressure and 4,150 rpm. Lundquist expected to obtain about 800 hp with the new cams. He also reported having run the unit at 5,000 rpm for ten minutes.

Problems had been encountered with the Scintilla magnetos from the R-2600. Point spring tension of 15-16 pounds was insufficient, allowing the points to bounce at high rpm. New points with 32 pounds of tension were to be tried.[11]

After the 50-Hour Endurance Test was completed, the engine was in good condition except for power shaft bushings, a camshaft bushing and intermediate power shaft bushing. Changes recommended to address problems found included the following: distributing oil directly from the pressure pump to the power shafts; metering the flow to the rear section; increasing clearances of the front and rear high-speed power shaft bushings; lead-indium plating of the master rod bearings and the front high-speed power shaft bushings; and installing floating bushings in the intermediate power shafts.[12]

Work Progresses on the T-42

Lundquist reported that the T-42 was in final stages of fabrication. Assembly would be complete in two weeks and the engine would be running in six. Both General Electric and Bendix Scintilla magnetos were being adapted for the R-2160.

Problems with weight were reported. The bronze cylinder banks had brought the total weight of the engine from 2,735 up to about 3,000 pounds. There was talk of etching away excess metal to reduce the weight of the castings. Consideration was given to holding the first R-2160 at Wright Aero until after the beginning of 1941. By having a shorter cycle between testing and changes, Wright Aero hoped the engine development process would proceed more rapidly.[13]

The First "Official" Run of the T-42

On November 4, 1940, then-Major F.O. Carroll, Chief of the Experimental Engineering Section at Wright Field, was on hand for the first run of the T-42. In a November 6 interoffice memorandum to the Chief of the Power Plant Laboratory, he reported that the engine had been run for about one hour, and had then been disassembled and inspected to make sure all moving parts were being properly lubricated. Several minor parts required enlarged oil passages, and several bushings were scored due to inadequate end clearance. Otherwise the engine was in good shape, and everyone seemed well pleased with the results of the tests. His glowing report concluded, "Some day the Tornado will be a beautiful engine for pursuit type airplanes. When viewed along side the R-3350, the Tornado seems to have about half the frontal area of the R-3350."[14]

Not mentioned in any of the official correspondence is a story that surfaced much later. E. K. "Spike" McCord responded to a Winter 1990 issue of *Propwash*, a quarterly newsletter published by the association of Curtiss-Wright retirees, about the Tornado refurbishment. He wrote to say that he had worked on the engine in Paterson at the time it was built and tested. "The first engine was tested on the X-1 experimental dynamometer test stand with some surprising results. The engine tester thought he had the throttle cracked. Actually it was 95% open (the throttle was hooked up backwards). When he turned on the ignition the 42 cylinders fired (as they were supposed to). But with no preload and wide open throttle the engine literally exploded inside, master rods, baby [articulated] rods, etc., broke and destroyed the engine. I believe it was the No. 1 engine. I can't remember the engine tester's name. Andy Dodd was there but I don't think he was running X-1 at that time."[15]

Jerry Abbamont, a Wright Aero test engineer, confirmed the story, as did John Minnich, another test engineer. Abbamont was in the control room with Andy Dodd, an observer on the test crew, when the incident occurred. "Everything you stated is actual. I remember the engine being motored over for the start and the ignition being turned on. There was just a very loud roar as the engine accelerated from starting speed through infinity. As the take-off speed was 4,150 rpm normally, it is anyone's guess as to what rpm that sucker seized at running against the unloaded dynamometer. I was a party to the draining of the sumps and distinctly remember there were no chips, and no oil either."[16]

The Army Orders Additional T-42s

In late 1940, five T-42s were being constructed in Wright's experimental shops. Two were intended for loan to Howard Hughes and were to be ready for flight about July 1941; one was scheduled for the Power Plant Lab; the other two would remain at Wright for development testing. Carroll was of the opinion that this was an insufficient number of engines. He warned that in order to get engines within one year, additional orders would have to be placed soon. He went on to voice the concerns of Wright Aero:

"The Wright Company seems to be quite concerned that the Army has expressed so little interest in this engine in the form of contracts. They suggest an immediate order for five additional engines, and would like the Army to plan on experimental installations in one single engine pursuit and one two engine pursuit project. It was explained to Wright Aero that whereas the engines for our single engine pursuit projects for this fiscal year were fairly definitely settled, that there exists a possibility of installing two Tornadoes in the Lockheed XP-58 project, provided Lockheed's design studies indicated that the application be suitable and the performance guarantee adequate." Carroll recommended purchase of an additional five engines, if Lockheed accepted the Tornado.[17]

On November 27, 1940, AFP No. 172797 initiated the procurement of three right-hand rotating R-2160-3 engines and two left-hand rotating R-2160-5 engines under contract W535 AC-13709. One R-2160-3 and one R-2160-5 were required for installation in the Lockheed XP-58 airplane. An R-2160-3 and an R-2160-5 engine were required for spares as well. The third R-2160-3 engine was to be used for tests in the Power Plant Laboratory.[18]

On January 13, 1941, Wright Aero requested an amendment to W535 AC-13709 changing the propeller rpm at take-off from 945 to 1040. This change delayed the projected shipment of the engine until June 15, 1941, but resulted in no cost increase.[19] This change, probably for the Lockheed XP-58, was needed in order to keep the propeller diameter small while still absorbing take-off power. Even so, the propellers were still huge, 14'8" diameter, four-blade models. Change Order No. 1 to AC-13709 was approved by the Under Secretary of War on April 24, 1941.[20]

The First Hint of Real Trouble, and the Schedule Starts to Slip

A February 18, 1941, letter from P.B. Taylor to the Materiel Division is telling. Taylor was then the chief engineer of Wright Aero. The letter withdraws a bid for the R-2170, a smaller version of the R-2600. It asserts two factors that were to severely hamper the Tornado project. First, Wright Aero lacked a dedicated experimental machine shop. Instead, it used production facilities and outside vendors to construct experimental engines.[21] This policy led to extreme delays in the development of experimental engines. Second, it hints that engineering and design resources are stretched to the limit. However, this was only the beginning of a host of engineering and management problems that would plague the company for the next five years.

Wright Aero announced the following delivery schedule to the Army on May 3, 1941:

Engine	Contract	Model	Allocation	Delivery
11766	Pending	R-2160-3	WAC Development	6-15-41
11767	Pending	R-2160-3	WAC Development	7-22-41
11768	13709	R-2160-1	Wright FieldTest	8-15-41
11769	Pending	R-2160-3	Dummy for XP-58	10-1-41
11772	Pending	R-2160-3	XP-58 Flight Engine	3-1-42
11769	Pending	R-2160-5	XP-58 Flight Engine	4-1-42
11773	Pending	R-2160-5	Spare	5-15-42
11774	Pending		Spare	7-1-42
11775	Pending		Spare	8-15-42
11776	Pending		Spare	10-1-42
11777	Pending		Spare	11-15-42
11778	Pending		Spare	1-1-43

Wright Aero wanted to retain engines 11766 and 11767 for development tests in Paterson where it had freedom to alter the engines as necessary as the tests proceeded. These engines were to be modernized or replaced before delivery to the Army, with the price being negotiated at that time. Similarly, in order to expedite progress on the Lockheed XP-58, engine 11769 was to be initially supplied as a dummy for use in installation engineering and later modernized and returned to Lockheed as a flight engine by April of 1942.[22] It should be noted that several other engines had already been produced. Engines 11760 and 11761, both 14-cylinder test units, were undergoing development testing at Wright Aero, as were the first full-sized T-42 engines, 11764 and 11765.

In order to streamline and consolidate the many Tornado procurements under consideration, new AFPs were issued on May 15, 1941. AFP No.178967 of Jan 23, 1941, which cancelled and superseded AFP No. 172797 of November 27, 1940, was combined with AFP No. 186038 to result in contract W535 AC-19438. AFP No. 178967 was for the amount of $457,000, and AFP No. 186038 in the amount of $172,500. Each of the prior AFPs covered procurement of three R-2160 engines with right-hand propeller rotation and two with left-hand propeller rotation, a total of ten engines.[23]

On June 20, 1941, Chief of Materiel Division transmitted to Under Secretary of War for approval contract W535 AC-19438 which covered procurement of five R-2160-3 and five R-2160-5 engines at the unit cost of $630,000. It was stated that three R-2160-5 and two R-2160-3 engines were needed for experimental test engines for expediting the development of the Tornado. Later, it was expected that the engines would be used in conjunction with experimental airplane development. The remaining engines were intended for use as Government Furnished Equipment on experimental planes. One of each type was required as a spare, and the other R-2160-3 was required for test purposes.[24]

During a June 18, 1941, conference at Wright Aero, difficulties with the R-2160 on the test stand were discussed. While operating at high horsepower and rpm, the engine had a tendency to lope. Lundquist believed this was due to poor fuel distribution. Wright Aero was researching various diffuser entrance channels and numerous types of gasoline sprays or spinners in front of the impeller, hoping to develop a method of breaking up the fuel/air mixture in a more proportional manner. It was believed that excess fuel, dropping to the bottom of the diffuser housing, was being picked up by the impeller, and causing a transient rich mixture and the ensuing lope. Rapid resolution of the distribution troubles was anticipated.

Fuel injection was being studied, and a mounting pad for a fuel injector pump was included at the front of each cylinder block.

Backfire screens were being considered, although no backfire problems had arisen. Three types of manifolds and screens were under consideration. One unit had been fabricated. It included one backfire screen about one-half thick and two inches wide, which, as in the Rolls-Royce Merlin, ran the length of the six intake ports.

Materiel Division asked if Wright Aero had experienced trouble with the 14-mm long reach mica-insulated plugs. Lundquist stated that no problems had been encountered in 201 hours for No.11760, 347 hours for 11761, 25:15 hours for 11765 and 18:50 for 11765 with aluminum blocks. Wright was informed of troubles with experimental 14-mm ceramic plugs in other applications.

Problems with the Tornado weight budget were being addressed. The second engine with aluminum blocks weighed about 2,703 pounds. Numerous changes were anticipated, including the use of magnesium parts. Backfire screens were expected to add about forty pounds.

A test rig belonging to Howard Hughes was to be shipped to the Wright Aero Flight Test Facility at Caldwell, New Jersey on July 15, 1941, and was expected to be operating with the Tornado installed by August 15. It was anticipated this test rig would be very useful in determining overall Tornado heat rejection figures. Oil flow of 300 pounds per minute with an inlet temperature of 180° F and an outlet temperature of 250° F was placing large demands on the oil cooler. It was hoped that by increasing inlet oil temperature to 200 – 210° F without changing the upper limit would decrease the heat rejection.

The first engine was to be shipped to Howard Hughes around December 1, 1941. The outside shell of an R-2160 was to be shipped to Lockheed in October of 1941 for installation in the XP-58. These shell engines were to be returned to Wright Aero, rebuilt, and returned to Lockheed on April 1, 1942. Engine No.11768 was to be delivered to Wright Field about August 15, 1941.[25]

The Trouble with Spark Plugs

At a July 9, 1941, conference in Paterson between Wright Aero and the Army, the possibility of using the same 14-mm short reach spark plugs that were installed in the Rolls-Royce Merlin was first discussed. Apparently, the Army wanted to standardize its spark plug inventory and avoid the use of the special long-reach plugs Wright Aero had chosen for the Tornado. At that time, Wright Aero pointed out two problems with the idea. First, the external dimensions of the 13/16" hex portion of the proposed spark plug would not fit into the counterbore in the Tornado cylinder head. Second, the short reach plug produced an undesirable shape to the cylinder dome. The small diameter of the Tornado precluded any easy change to accommodate the larger plug. Over the next sixteen months, numerous pieces of correspondence were traded between the Army, Wright Aero, AC Spark Plug Company, and Champion Spark Plug Company. The Tornado was ultimately fitted with Bendix BG-466-S spark plugs, as had been originally specified.[60]

On July 14, 1941, at a conference in Gen. Henry H. "Hap" Arnold's office, Materiel Division was directed to "try to get a Tornado engine installation by next July 4th, 1942." On July 25, 1941, Brig. Gen. O. P. Echols informed General Arnold that it was doubtful whether the Tornado would be ready for installation by that time. He further stated that the delivery schedule provided for the first two engines to be retained at Wright Aero. A third was slated for shipment to Wright Field about August 15, 1941. A dummy engine was to be sent to Lockheed about October 1, 1941. The next two production engines were scheduled for shipment to Lockheed on March 1, 1942, and April 11, 1942, for installation in the XP-58. The two T-14s had accumulated 535 hours and the two T-42s, 44 hours, all at Wright Aero. Bell Aircraft had made preliminary studies of installing a Tornado in the P-39, but shaft difficulties were expected to preclude this.[26]

While working with Lockheed on the XP-58 installation, Wright Aero developed a more efficient engine installation, which reduced the installed weight of each engine by about 25 pounds at the expense of a 0.75-inch length increase. This was reported to Materiel Division on August 5, 1941. The length increase was not perceived to be significant.

In order to effect the improved installation, three changes were made. The steel mounting plate was redesigned to take the entire sheer load and to be located between the crankcase main section and supercharger front housing. A new coolant pump and supercharger rear housing were designed to relocate coolant connections, thereby improving accessibility to the two lower engine mounting plates. The rear main oil pump was redesigned to provide a more compact unit with the same capacity. The new pump also had two scavenge outlets that made possible the use of parallel oil cooler installations. The front oil scavenge pump was redesigned to incorporate the same features as the rear pump. Wright Aero recommended the incorporation of these features in the last three R-2160-3s and all R-2160-5s at no cost to the Government and with no program delay. The remaining two R-2160-3s at Wright Aero would be fitted with the new oil pump design for development testing.[27]

On August 23, 1941, Wright Aero was directed to deliver the engine mock-up that had been loaned to Hughes Aircraft to Vultee for use in its XP-54 program. The Army anticipated Hughes would be done with installation engineering by September 15, 1941, and that the engine should be at Vultee by October 15, 1941.[28]

During the last week of August 1941, a T-14 engine, 11760, was busy testing different cam profiles in an attempt to increase power by earlier opening of exhaust valves. By September 3, 1941, this engine had logged 274:20 hours. Meanwhile, a

White Lies and the R-4090

Despite the delays that had begun creeping into the Tornado program, and despite its complaints about a lack of experimental machining capacity, Wright Aero had, during the first half of 1941, been secretly trying to sell a 3,000 hp air-cooled 22-cylinder engine to the Navy. When Dr. Arthur Nutt, Wright Aero's vice president of engineering was questioned by the Army about this on July 22, 1941, he said that Wright Aero had assumed that, with the Tornado project under way, the Army would not be interested in a 3,000 hp air-cooled engine. He also implied that the Navy had the project under wraps, preventing him from discussing details, and stated that, "he wished the Army and Navy would get together on this project so that he would be free to discuss all details."[61]

The Navy replied that Dr. Nutt had proposed the development, had requested the confidential classification, had not yet produced a proposal, and that the Navy had no objection to Army having all available information concerning the proposed development.[62] After this rocky start was brushed aside, a joint development of the 3,000 hp R-4090 was begun. The project continued at least into 1944. It is thought that two copies of the engine were built and run. However, the project was eventually abandoned. At a time when both the Tornado and R-3350 programs were in serious trouble, when managers, engineers, and technicians were in very short supply, Wright Aero was trying to perfect a new dead-end project instead of honoring its existing commitments.

The air-cooled R-4090, using twenty-two R-3350 cylinders in two rows, was to have been competition for the Pratt & Whitney R-4360. (Courtesy of the New Jersey Aviation Hall of Fame and Museum)

second T-14, engine 11761, was in the midst of a second 150-hour test. Problems with the master rod bearings were delaying the test. This engine had logged 396:50 hours.

T-42 engine 11764 was encountering problems as well. A test on August 25 of an air baffle that was to prevent heating of the intake charge by coolant passages had not resulted in a power increase. When the engine was torn down, cracks were discovered in two cylinder banks, supercharger impeller spacers were broken, and the impeller had rubbed against the supercharger front housing. Additionally, the propeller shaft bearing journal, located inside the front crankshaft main bearing, was badly worn. Cylinder block failures had been a frequent problem. Most of the cracks had arisen on the sides of the cylinder castings and had then propagated into the intake and exhaust valve ports. Casting porosity and core shifts were blamed. The patterns were remade to incorporate thicker cylinder block walls. Total time on engine 11764 was 50:30 hours. Engine 11765 was being rebuilt. It had logged a total time of 35:45 hours.[29]

In an attempt to cure the power output problems of T-14 engine 11760, intake manifolds were altered. This engine had been rapidly assembled from available parts that included a diffuser section from the R-2600. Initially, intake manifolds issued from every other diffuser outlet with the alternate holes plugged. This was changed so that the intake manifold for each block of cylinders was fed by two adjacent holes. Exhaust ports were being enlarged on T-42 engine 11764 in hopes of achieving increased power output. Engine 11765 had been fitted with new intake manifolds that were divided longitudinally, making each one, in effect, two manifolds. One half supplied the odd-numbered cylinders, and the other, the even numbered cylinders.[30]

What transpired during the following month is a matter of speculation. Three progress reports are missing. When the record resumes on October 7, 1941, T-14 engine 11760 is back in experimental assembly having cylinder block hold down studs replaced. There is also a problem with the record of

Much Ado about Nothing—XP-58 Cooling Problems

In early April of 1942, "another very serious problem" came to light. Lockheed's XP-58 design had been predicated on an engine with 2,350 hp with a heat rejection of 35 percent running a coolant temperature of 250°F with an outside air temperature of 100°F. Wright Aero had increased the heat rejection figure from 35 to 40 percent. This meant that the XP-58 cooling system was no longer adequate, and that the aircraft would have to be redesigned with larger radiators (with the corresponding schedule slip), flown at reduced performance, or with a higher coolant temperature of 260°F. From the point of view of Lockheed and Material Division, the higher coolant temperature was the only viable option, and Wright Aero was to notify Materiel Division immediately of its decision.[63]

In a May 23, 1942, memorandum to the Chief of the Power Plant Laboratory, Col. F. O. Carroll recapped the heat rejection issues raised in the April 6 letter to Wright Aero. He recounted that Wright Aero was asked to run the type test at a coolant temperature of 260°F instead of 250°F. "To this question they have remained silent. The local representative, Mr. Harvey Holland, has advised this office that recent information indicates that the heat rejection will be less that 40 percent and will be closer to 35 percent. This is unofficial and can be taken for what it is worth." If the heat rejection turned out to be 37 percent, the cooling system on the XP-58 would have been marginal but, in actual practice, satisfactory due to the thermal inertia of the system.[64]

On September 2, 1942, Wright Aero finally answered the mail in the Lockheed XP-58 heat rejection issue. After calibration of the test equipment, data indicated that at military power, coolant heat rejection was just under 35 percent and that oil cooler heat rejection was just under 14 percent. This was below the 50 percent figure that had been previously set as an absolute maximum. All of this was determined with a 250°F coolant temperature, which was the value used by Lockheed and Republic in designing aircraft coolant radiators and plumbing.[65] As a part of the heat rejection testing, Wright Aero had also tested various mixtures of water and ethylene glycol. It was discovered that the coolant mixtures afforded radiator skin temperatures that were 40°F to 50°F higher than with pure ethylene glycol. Wright Aero pointed out the advantages of smaller radiators and reduced aircraft drag and urged the adoption of a 60/40 mix of ethylene glycol and water, as well as the elimination of the Tornado research dedicated to pure glycol.[66] Lockheed and Republic were consulted and were both in favor of elimination of the pure ethylene glycol requirement. The change was approved.[67]

Republic aviation was also informed of the new heat rejection data. Oil flows of 350 pounds per minute were assumed at the maximum allowable oil inlet temperature of 220°F, and 325 pounds per minute at the more reasonable 185°F for cruise operation. While formal type test specifications prohibited changes to the 185°F value, Wright Aero had decided that the 250°F value was acceptable for takeoff, military power, and ground operations. A deviation in the aircraft specification to reflect this higher "emergency" oil temperature value was suggested.[68]

time on the engine. It is now listed as 245:20, 25 hours less than the prior report. Engine 11761 was finally being assembled after its long wait for master rod bearings. Engine 11764 had achieved a maximum power of 2,350 hp at 4,410 rpm on September 30. However, this came at a price. There had been carburetor and ignition trouble that resulted in the replacement of the carburetor along with numerous spark plugs in "C" bank. A failure of the E-6 cylinder was traced to the manifold vane between cylinders 4 and 5 having broken off and been ingested by cylinder 6. Eighteen hours had been accumulated on the engine during the period, for a total of 68:30 hours. What transpired with engine 11765 is not known. It was being rebuilt to take the place of the wrecked 11764 and by October 7, 1941, had logged 72:20 hours.[31]

The Successful 150-Hour Endurance Test of the T-14

Wright Aero published the results of T-14 tests on November 11, 1941. These tests were conducted in Paterson from March 1 through July 14, 1941. The summary stated:

The engine operation and performance during the 150-hour test were satisfactory considering that the test was of a development nature. No difficulty was encountered in securing the powers and speeds required and control of the engine was good over the entire range. Some improvements were necessary as was shown by parts breakage. Some of these improvements had already been tested and others were ready for testing. No fault was found for which there was no readily available improvement. Major troubles were few when the shortness of the development period on the engine is noted. The good condition of the engine at the completion of the 150 hours of endurance is evidence of its durability except for scattered correctable faults.[32]

The Second World War Changes Priorities

At this point in time, the outlook for the Tornado looked rosy. While there had been setbacks, challenges, and schedule slips, and while there were still problems to be addressed, development of the engine was progressing more or less normally. Then came December 7, 1941. Subsequent U.S. entry into World War II imposed huge changes on Wright Aero. The biggest was the increased priority of the XB-29 and the Wright R-3350 that was to power it. In order to drop bombs on Japan by June of 1944, engines were desperately needed. But R-3350 development was going badly. Additionally, Wright Aero was challenged to produce enough R-1820s and was having quality problems with the R-2600. Wright Aero was forced to throw all its engineering talent at these projects.[33]

A new allocation and delivery schedule for R-2160s was decided at a February 14, 1942 conference at Wright Field:

Engine	Allocation	New Delivery	Old Delivery
11767	Type Test	8-1-42	12/41
11768	XP-58 or DX-2	9-15-42	3/42
11769	XP-58 or DX-2	11-15-42	5/42
11770	XP-58 or DX-2	1-15-43	5/42
11771	XP-58 or DX-2	3-1-43	?

Additionally, Arthur Nutt of Wright Aero had said it was mandatory that engine No. 11766, initially scheduled to be the first installation engine, be kept at Wright Aero to expedite development of satisfactory flight engines.[34]

On March 25, 1942, Wright Aero advised Materiel Division that "as a result of this company's not being able to secure during the past year sufficient parts and labor skilled in experimental work, and due to the higher priority of other projects, it has been necessary for this company to delay shipment of the Model R-2160-1 engine on subject contract until August 1942." The schedule also slipped for the additional R-2160-3s and –5s. Wright Aero further observed that since the single R-2160-1 was scheduled for a 50-hour military approval test, it would make sense to ground test an engine identical to those which would be flown. Wright recommended that the R-2160-1 be assembled as an R-2160-3 instead, in accordance with Wright Aero Engine Specification No. 689-D.[35]

Wright Aero, in a March 31, 1942, letter to Lt. Col. M. S. Roth, related a disagreement over promises of rated power for the R-2160. Republic Aircraft was developing the XP-69 and counting on 2,500 hp from the Tornado to achieve the aircraft's promised performance. Wright Aero agreed that 2,500 hp had been an eventual goal but had never been promised to Republic for their first engines. They asked for Wright Field's "informal" assistance in smoothing out this wrinkle.[36] By April 6, 1942, the issue had been pushed to the level of the Chief of the Materiel Division. Col. F. O. Carroll agreed with Wright Aero about the initial promise of 2,500 hp and even stated that the specification for the XP-69 called for installation of an engine with a military rating of 2,450 hp. But he also was very clear that "The development of this engine to 2,500 horsepower should not be neglected. Every effort should be made on the part of the Wright Aeronautical Corporation to obtain this 2,500 horsepower rating on engines contemplated for production."[37]

Experimental Engineering Section informed General Echols on April 21, 1942, that diversion of Wright Aero experimental production facilities to the R-3350 program had caused further delays in the Tornado program. It became obvious that Wright Aero would be unable to supply engines to both aircraft programs. There was talk of a second XP-58, and the Republic XP-69 was awaiting an engine as well. It was decided that R-2160s promised to Howard Hughes for the DX-2 would be diverted to the XP-58 project. Engineering Division made arrangements for Hughes to receive three Pratt & Whitney R-2800-49 engines instead.[38]

On June 22, 1942, a conference was held at Wright Aero and again conferees pushed Tornado delivery dates to the right, stating that the first R-2160-3 would not be available until September 1942 at the earliest. Delivery of the first XP-58 engine would be delayed until November. Still more personnel and testing facilities had been diverted to the R-3350 program. The R-2160 had also suffered some problems. These included cylinder block cracks, connecting rod failures, valve failures, distribution difficulties, and ignition difficulties.

Solutions to all of these problems were in process but were moving slowly. The cylinder banks had been redesigned to strengthen the exhaust port area, but none were yet available for test. The connecting rod failures were due to liquid in the combustion chambers. This was to be addressed by a revised priming system. Other connecting rod failures had occurred because of excessive side clearance in the piston pin bosses, but the cause of this was still under investigation. Valve failures due to valve spring resonance and subsequent release of the valve locks were being addressed by increasing the cone length on the washer and by use of a circlet to hold the locks in place. Another cause of valve failure was interference with the pistons at high speed because of valve spring resonance causing slow valve closure. Redesign of the valve springs was expected to address both types of valve failures. Four different new intake manifold designs were being tested to solve the distribution difficulties. The ignition problems were caused by spark plugs with mica insulators. Wright Aero thought it might need Materiel Center assistance in obtaining ceramic spark plugs of a special 14-mm size.

There were also delays in ground testing on the borrowed Hughes test rig at Caldwell, New Jersey. This was due to propeller spinner and fan tests still under way at Wright Field. Little progress had been made on the coaxial propeller drive for the XP-69, still in the design stage. Development of a similar design for the R-3350-17 was expected to be useful in the Tornado application.[39] While at first glance the list of problems appears daunting, it must be remembered that all T-42s in existence had logged less than 700 hours. For an engine program with so little running time, the problems were minor and would have undoubtedly been ironed out with more work. Lack of personnel, facilities and motivation were the real cause for such slow progress in the Tornado program.

Reports on development progress of the Tornado again became spotty. This was probably due to several factors. The primary reason was that they simply did not get into the file and consequently were lost. There were also program changes that could offer explanation as well. H. A. Damron, author of the earlier reports, at some point was replaced by W. E. Bettoney. There must have been some problem with Mr. Bettoney's reporting, because he was ordered by a letter from Wright Field to submit prompt weekly progress reports. On February 12, 1942, Major Belt chides, "The latest progress report on file in this office on the above subject [R-2160] is No. 28 dated January 30, 1942. It is requested that Mr. Bettoney be instructed to prepare at least one report on this subject each week and to forward one such report to the Materiel Division in time for it to be received in this office during the first part of the week, not later than Wednesday Afternoon."[40] No further reports exist in the file, however, until Report No. 50 dated July 25, 1942.

At that time, work on the T-14 engines had essentially ceased. Four new T-42 engines had been added to the stable of test engines – numbers 11766 through 11769. Engine 11764 had been fitted with constant velocity intake manifolds and provisions for increased crankcase breathing. Several thermocouples had been installed on the exhaust valve inserts. This engine had logged 240:40 hours as of July 25, 1942. Engine 11765 had been equipped with a new high-turbulence piston. Since no single-cylinder testing of the Tornado series had ever been contemplated, this new piston design was first tested in the complete engine, with disastrous results. One of the pistons had broken through the piston pin boss during run-in. This failure precipitated a broken connecting rod that wrecked the engine. Total time on 11765 was now at 325:25 hours. Engine 11766 was supposed to be run on the test rig provided by Howard Hughes at Caldwell Airport, New Jersey. This testing had been delayed, however, because it required the special propeller spinner that was still undergoing development at Wright Field. Time on engine 11766 was 79:25 hours. Engine 11767, with a total time of 36:20 hours, was having its pistons reworked by drilling radial oil drain holes to provide better oil control. Engine 11768 had been used as a mock-up to prepare test stand X-32 for Tornado engines. Its total time was 60:45 hours. Engine 11769 was being assembled. Total time on all T-42 engines stood at 742:50 hours.[41]

Two weeks later, engine 11764 had been removed from the test stand after having completed a round of tests. The engine was low in power and burning oil at the rate of about forty pounds per hour. It was believed the excessive oil consumption was due to the conventional ring arrangement. Time was 277:05 hours. Engine 11765 had been run-in. Oil consumption with the new ring setup was reduced to eight pounds per hour. This engine had been built with the high turbulence pistons, which had presumably been fixed. Total time was 332:35 hours. The remainder of the engines had been inactive except that 11767 had been involved in a test to see if a connecting rod could be bent with the starter and a cylinder filled with oil. Despite having stopped the starter, no connecting rod damage was incurred.[42]

Testing progressed through August of 1942 with the usual mishaps. Engine 11764 had been fitted with strain gages under the camshaft. While these gave good indicator diagrams, calibration errors resulted in incorrect peak pressures. Running had been interrupted by a broken valve washer and by a seized rear oil pump drive gear. Time stood at 284:00 hours. Engine 11765 had been run as fast as 4,400 rpm, but a broken supercharger thrust washer ended the test. Efforts to eliminate backfiring by turning the pistons around and installing a third valve spring on each valve were not effective. No power increase resulted from the use of high turbulence pistons. Time was 432:35 hours. Engine 11766 was being reassembled after bent connecting rods resulted from a hydraulic lock. The team planned a new priming system and installation of manifold drains. Time was 79:25 hours. Engine 11767, with 36:20 hours, was being assembled with special

What was The Real Reason for Connecting Rod Breakage?

Throughout the development of the T-42, the design team was vexed with occasional breakage of connecting rods. This was blamed on excessive priming and improper starting procedures. Although much experimentation was done to identify the root cause of these problems, no "smoking gun" was ever found.

While excessive priming may have been a partial contributor to engine damage, the real cause may have been a much more insidious one. As anyone who has worked around radial engines knows, oil leaks past the piston rings and into the lower cylinders when a radial engine sits idle for more than 30 minutes. In some cases, enough oil enters a lower cylinder to completely fill it, causing a hydraulic lock and preventing rotation of the engine. Two approaches were used to address this problem in conventional radials. First, the ground crew would "pull the prop through" at least two full revolutions of the engine to pump any residual oil out of the cylinders and into the exhaust where it would be harmlessly expelled. Second, the starter was equipped with a torque-limiting clutch that would slip if a hydraulic lock prevented free rotation of the engine. If a hydraulic lock was encountered, the lower spark plugs were removed and the engine turned over to expel trapped oil from the lower cylinders. With the Tornado, access to spark plugs was difficult, so the operators may have resorted to other, easier means for clearing hydraulic locks.

Another technique was used by some uninformed crews to clear hydraulic locks. This involved turning the prop backwards, which forced the slug of oil into an intake manifold. While this technique cleared the hydraulic lock, the oil in the manifold was a potential time bomb. When the engine was started, the slug of oil could be ingested into the cylinder and again cause a hydraulic lock. When this happened to a running engine, the results were always disastrous, producing bent connecting rods, broken pistons, and damaged cylinder heads. To make matters worse, the damage might not be immediately obvious. A bent rod might run for many hours before breaking, long after the offending hydraulic lock had occurred. Still, it literally took decades to get the word out to most radial engine operators that turning the prop backwards was a bad idea. Indeed, the practice persists in some circles even today.

In the Tornado, intake manifolds for cylinder banks "C," "D," and "E" were all at least partially below the level of the cylinder. Oil could have entered the intake manifolds through an open intake valve, or by careless handling by the engine operators. In either case, it is this author's opinion that a slug of oil ingested from the intake manifold could have accounted for all the trouble with broken rods and pistons. The problem could have been easily solved by the addition of a simple manifold drain arrangement for the three lower cylinder banks.

master rod end spacers. Engine 11768 was also getting the same spacers. A cylinder block had previously broken out, but no explanation was available. Time was 76:15. Engine 11769 was still being assembled. A T-14, engine 11761, was being reworked for static testing of connecting rods. Total time on the T-42 engines was now at 818:35 hours. A new starting procedure had been developed to prevent hydraulic lock.[43]

The first week of September brought a new reporting engineer, 2nd Lt. Robert L. Richardson, and the usual troubles with engine 11764. Several spark plugs were replaced before a heat rejection run could be completed. This run was made with an oil inlet temperature of 185°F at an engine power of 2,000 hp. Unfortunately, the wrong thermometer readings were taken and the results were useless. Further insult came from a stretched G-5 intake valve and a broken F-6 valve washer that allowed the valve to drop into the cylinder and damage the piston. Total time on 11764 then stood at 296:45 hours.

The reduction gear support on engine 11765 had been reworked to add 0.015 inches to the pinion bushing depth. This was done to give the pinions more float. Time was 342:35. Engine No. 11766 was being installed on the Hughes test rig, and had logged 79:25 hours. Engine 11767 with 45:50 hours was being installed on a test stand for a 150-hour test at a normal rated power of 1,800 hp at 3,800 rpm and takeoff power of 2,000 hp at 4,000 rpm. A new cylinder block was finally ready for 11768. Engine 11769 was awaiting parts for the rear section and reduction gear, but completion was expected during the second week of September.

Engineers working on engine 11761, after filling a cylinder with heavy oil, had finally succeeded in bending a connecting rod during static tests. A lack of test stand and technician time continued to impede progress with the Tornado.[44] Lt. Richardson's report must have been late, because it inspired a letter from Wright Field demanding weekly progress reports that arrived no later than Tuesday. The letter even went so far as to suggest that the report be prepared and forwarded on Saturday in order to ensure making the deadline.[45]

The threat from Wright Field must have been effective because the next report, late as usual, comes from Ernest C. Saltzman. Engine 11764 remained disassembled to install special crankcase drains for improved scavenging. Engine 11765 was still trying to get through a heat rejection test but was troubled by a burned out rear impeller thrust washer, a stretched A-6 intake valve, and collapsed heads on exhaust valves A-5, A-6 and E-1. The engine had logged a total of 351:10 hours. At Caldwell, New Jersey, engine 11766 was now running with the newly developed propeller spinner, but was having oil and coolant cooling difficulties. Operation without the spinner was contemplated while the aerodynamic characteristics of the spinner were further investigated. Engine 11766 now had accumulated 81:25 hours. Engine 11767 was being rebuilt with silver master rod bearings, replacing the steel-backed copper-lead ones used until then. It showed a total time of 59:10 hours. Engine 11768 had completed six hours of the 150-hour endurance test. It had logged a total of 95:35 hours. Engine 11769 had finally completed its six hour green run. Since parts of engine 11770 were not available, major components of 11769 were assembled as a mock-up engine and forwarded to Wright Field. Total time was 6:20 hours. Total time on all engines was 890.25.[46]

A September 21, 1942 telegram to Wright Aero from Materiel Center requested express shipment of a display Tornado engine with turbosupercharger.[47] This was to be returned within one month. Since parts of engine 11770 were not available, major components of 11769 were assembled as a mock-up engine and forwarded to Wright Field.[48] These items were returned to Wright Aero, along with an R-3350-13 that had also been borrowed, on October 10, 1942.[49] These items were part of an engine exhibit that had been assembled at Wright Field on October 8 and 9 for the review or Gen. O. P. Echols, Commanding General, Materiel Command.

A conference was held on October 17, 1942 between Mr. Ray Young of Wright Aero and Materiel Center Representatives Col. E. R. Page and Mr. O. P. Chenoweth. Young explained that Tornado development had been delayed because of lack of test equipment. The Tornado demonstration engine that had appeared at Wright Field on October 8 and 9 had captured the interest of the AAF Materiel Command. He ordered that three Tornado test stands be built at Caldwell. These were to be completed around November 15 for use in endurance testing.

Materiel Command had heard that Wright Aero was increasing the diameter of the engine. Young explained that while the 2,350 hp version would remain the same size, that 2,500 and 2,700 hp versions would be around 40" in diameter. The

Tornado brake mean effective pressure per inch of manifold pressure was about 75 percent of what was being achieved in the Wright G-200 Cyclone (R-1820). The increased diameter would result in improved combustion chamber design, improved porting, and standard spark plugs.[50]

Wright Aero Slips the Schedule Yet Again, and the Axe Falls

In a February 9, 1943 telegram, Engineering Division notified the Assistant Chief of Staff that "Wright Aeronautical Corporation has informed Lockheed that delivery dates of Tornado Engines for installation in the XP-58 airplane are indefinite. In view of the fact that the airplane has been awaiting engines since last fall, Lockheed and we are most unhappy." Installation of Pratt & Whitney R-2800s had been considered, but the idea had been rejected because, even at War Emergency Power, R-2800s were not powerful enough. Allison V-3420s had also been studied and looked very good. In addition, the V-3420s were immediately available (February and March, 1943), as were suitable propellers. "We believe this is to be the only logical method of salvaging this project and consequently are proceeding as outlined above. This change is almost certain to adversely affect the Tornado engine picture. It looks to be just a metter [sic] of time until the Wright Company throws up the sponge as far as the Tornado engine is concerned anyway. We might as well face it now as later."[51]

Notification was received on May 11, 1943, that the contract for the Republic XP-69 would be terminated. Since there was no longer an application for the Tornado, Power Plant Laboratory was directed to determine the exact status of the contract with Wright Aero with a view to canceling it. Power plant laboratory was directed to procure only the Tornado engines it desired for development purposes.[52]

This activity had been completed by November 8, 1943. Specific attention had been given to design, development, calibration, and test programs of the R-2160 engine. It was concluded that the project did not look very promising in its present state, and in all probability will not come through in sufficient time to be of any use in the war effort. Moreover, no further airplane applications existed. Several other engines in the same power range as the Tornado were then in use. The R-2160 development program had been retarded due to concentration of engineering efforts on the R-3350 program, and this situation was not expected to change because of the priority established for the R-3350 in the B-29 program. It was recommended that the contract with Wright Aero for the R-2160 engine be terminated.

At that time, two contracts were active — Contract No. W535 AC 13709 dated January 23, 1940 and Contract No. W535 AC-19438 dated June 24, 1941. These covered procurement of eleven T-42 engines for a gross amount of $907,800. None of these engines had been delivered to Wright Field. Further, Wright Aero, because of low priority, was submitting no reports. [53]

A December 18, 1943, letter from T. M. Lucas, Wright Aero Director of Contract Administration to Commanding General, Materiel Command, outlined Wright Aero's position on contract termination:

> Six full scale engines have been built and operated a total of 1,500 [actually 1,631.35] hours, a considerable portion of this time being at 2,000 hp or better. (The contract requirement is 2,000 hp Normal and 2,350 hp Military Rating.) An additional thirteen sets of parts were fabricated and two 14-cylinder and one 6-cylinder test engine were built and run 1,200 hours and 130 hours, respectively, to obtain detail data.
>
> The above-mentioned testing indicated that the novel features of the engine were generally satisfactory but that the more conventional features such as intake manifolding, valve mechanism, and minor details were not satisfactory. Repeated trouble was experienced from "hydraulicking," but the most serious difficulty was lack of expected power because of an unsatisfactory induction system. It is believed that a fuel injection system might have cured some of the troubles, but the greater urgency of other projects in work for the Government has prevented further experiments.
>
> Inasmuch as the establishment of sufficient company priority to carry on the additional development work necessary to provide a satisfactory 2,350 hp engine would seriously conflict with other projects in work for the Government, and since the requirement tendency is toward engines of greater horsepower, it may be of benefit to the Government to cancel subject project.
>
> In the event the Government considers it desirable to cancel the subject project, this company will accept cancellation of the ten engines on Contract W535 AC-19438

without charge against that contract and will accept cancellation of the one engine on Contract W535 AC-13709 without charge against that contract other than the value of Items 4, 5, 6, and 7 which have been completed and delivered in accordance with the terms of the contract. These items and the values thereof are as follows:

Item 4 – One (1) crankcase, complete with bearings and crankshafts, for Model R-2160-1 aeronautical engine in accordance with Wright Aeronautical Model Specification No. 617-D with the exception noted in Item 2 of this article herein before referred to. $80,000.00

Item 5 – One (1) accessory case, complete with all accessory drives, integral supercharger and all Contractor-furnished accessories which are mounted on the accessory case, for Model R-2160-1 aeronautical engine in accordance with Wright Aeronautical Model Specification No. 617-D with the exception noted in Item 2 of this article herein before referred to. $40,000.00

Item 6 – One (1) reduction gear, complete, for Model R-2160-1 aeronautical engine in accordance with Wright Aeronautical Model Specification No. 617-D with the exception noted in Item 2 of this article herein before referred to. $10,000.00

Item 7 – One (1) set cylinder banks, complete with valve gear, pistons, connecting rod assemblies for Model R-2160-1 aeronautical engine in accordance with Wright Aeronautical Model Specification No. 617-D with the exception noted in Item 2 of this article herein before referred to. $90,000.00

Total $220,000.00

Attention is invited to the fact that the above-mentioned items do not comprise a complete engine. At the option of the Government, for an additional $10,000.00, this company will assemble and deliver to the Government a complete operatable [sic] engine in accordance with the requirements of Item 8 except that turbo-superchargers with related parts will not be furnished and no running acceptance test will be required.[54]

Contract W535 AC-19438 for the ten R-2160s was cancelled at no cost to the Government on February 12, 1944.[55]

The Army accepted Wright Aero's offer to assemble the materiel of Contract W535 AC-13709 and supplied the additional $10,000.00. The contract was cancelled at a cost of $230,000.00.

According to figures submitted by Wright Aero, the total cost of development had been $6,400,000[56]. This means that Wright Aero contributed $6,170,000 of its own money to the project.

A single R-2160-3, Serial Number 11770, was assembled and delivered to Wright Field. The date of this delivery is unknown. The remaining Tornado engines at Wright Aero were scrapped. A number of pistons and other small components wound up in the private collections of employees.

Reference Abbreviations

AFHRA – Air Force Historical Research Agency, Maxwell AFB, Alabama.
NARA – National Archive and Records Administration Branch Depository, Suitland, Maryland.
RD – A subgrouping of Record Group 342, the "Sarah Clark Collection"
RG – NARA Record Group Number
E – NARA Entry Number
B – NARA Box Number
AHoF – Aviation Hall of Fame and Museum of New Jersey

1 *Case History of the R-2160* - Brief.

2 T. M. Lucus, Manager – Order and Contracts Division, Wright Aeronautical Corporation, letter to Materiel Division, Army, December 27, 1939, *Case History of the R-2160* - File.

3 *Case History of the R-2160* - Brief.

4 Lundquist, *Tornado Design Studies*.

5 *Stress Analysis of R-2160*.

6 Hunter, *Supplement No. 1*.

7 O. P. Chenoweth, Army, Materiel Division Memorandum Report EXP-M-57-522-37, June 20, 1940, *Case History of the R-2160* - File.

8 Capt. O. F. Carlson, Army Materiel Division Memorandum Report EXP-M-57-503-184, August 10, 1940, *Case History of the R-2160* - File.

9 Ibid.

10 Routing and Record Sheet, Chief of Army to Chief, Materiel Division, August 30, 1940, *Case History of the R-2160*, - File.

11 Capt. O. F. Carlson, Army Material Division Memorandum Report EXP-M-57-503-224, October , 1940, *Case History of the R-2160*, - File.

12 Swords, *50 Hour Development Test*.

13 Capt. O. F. Carlson, Army Material Division Memorandum Report EXP-M-57-503-224, October , 1940, *Case History of the R-2160*, - File.

14 Major F. O. Carroll, Inter-office Memorandum from Chief, Experimental Section to Chief, Power Plant Laboratory, Wright Field, November 6, 1940, *Case History of the R-2160*, - File.

15 McCord, Letter to the Aviation Hall of Fame and Museum, February 13, 1990.

16 Abbamont, Letter to E. K. McCord, February 24, 1990.

17 Ibid.

[18] *Case History of the R-2160* - Brief.

[19] T. M. Lucus, letter from Wright Aeronautical Corporation to Assistant Chief, Materiel Division, Army, January 13, 1941, *Case History of the R-2160*, - File.

[20] Brig. Gen. Oliver P. Echols, Letter from Chief, Army to Assistant Secretary of War, April 23, 1941, and associated approvals dated April 24, 1941, *Case History of the R-2160*, - File.

[21] Taylor, Letter to Assistant Chief, Materiel Division, February 18, 1941.

[22] T. M. Lucas, Letter from Wright Aeronautical to Assistant Chief, Materiel Division, May 3, 1941. *Correspondence Regarding the R-2160.*

[23] Authority for Purchase No. 186038, May 15, 1941, *Case History of the R-2160*, - File.

[24] Brig. Gen. O. P. Echols, Letter from Chief, Materiel Division, Army to Under Secretary of War, June 20, 1941, *Case History of the R-2160*, - File.

[25] W. W. Woodward and O. F. Carlson, Memorandum Report EXP-M-57-503-390, July 2, 1941, *Case History of the R-2160*, - File.

[26] Routing and Record Sheet, July 14, 1941, *Case History of the R-2160*, - File..

[27] T. M. Lucas, Letter from Wright Aeronautical Corporation to Assistant Chief, Materiel Division, August 5, 1941, *Case History of the R-2160*, - File.

[28] F. O. Carroll, Letter to Wright Aeronautical Corporation, April 23,1941, *Correspondence Regarding the R-2160.*

[29] H. A. Damron, R-2160 Development Report No. 12, Serial Number 41-81, September 3, 1941, *Correspondence Regarding the R-2160.*

[30] H. A. Damron, R-2160 Development Report No. 13, Serial Number 41-86, September 9, 1941, *Correspondence Regarding the R-2160.*

[31] H. A. Damron, R-2160 Development Report No. 17, Serial Number 41-106, October 7, 1941, *Correspondence Regarding the R-2160.*

[32] Lundquist, *150 Hour Endurance Test.*

[33] Johnson, "Why the Boeing B-29 Bomber."

[34] Col. B. W. Chidlaw, Memorandum from Experimental Engineering Section, Wright Field to General Echols, April 21, 1942, *Case History of the R-2160*, - File.

[35] T. M. Lucas, Letter from Wright Aeronautical Corporation to Chief, Materiel Division, March 25, 1942, *Case History of the R-2160*, - File.

[36] Burnham Adams, Letter from Wright Aeronautical Corporation to Experimental Engineering Section of Materiel Division at Wright Field, March 31, 1942, *Case History of the R-2160*, - File.

[37] Col. F. O. Carroll, Letter from Chief, Experimental Engineering Section to Wright Aeronautical Corporation, April 6, 1942, *Case History of the R-2160*, - File.

[38] Col. B. W. Chidlaw, Memorandum from Experimental Engineering Section, Wright Field to General Echols, April 21, 1942, *Case History of the R-2160*, - File.

[39] Maj. M. C. Draler, Memorandum Report to Colonel Chidlaw, June 24, 1942, *Case History of the R-2160*, - File.

[40] Major Dean B. Belt, Letter to the Army Resident Representative at Wright Aeronautical Corporation. February 12, 1942 *Correspondence Regarding the R-2160.*

[41] W. E. Bettoney, R-2160 Development Report No. 50, Serial Number 42-170, July 25, 1942, *Correspondence Regarding the R-2160.*

[42] W. E. Bettoney, R-2160 Development Report No. 52, Serial Number 42-188, August 8, 1942, *Correspondence Regarding the R-2160.*

[43] W. E. Bettoney, R-2160 Development Report No. 54, Serial Number 42-193, August 22, 1942, *Correspondence Regarding the R-2160.*

[44] 2nd Lt. Robert. L. Richardson, R-2160 Development Report No. 56, Serial Number 42-205, September 5, 1942, *Correspondence Regarding the R-2160.*

[45] Major A. F. Wentzel, Letter to the Army Resident Representative at Wright Aeronautical Corporation. September 17, 1942 *Correspondence Regarding the R-2160.*

[46] Ernest C. Saltzman, R-2160 Development Report No. 58, Serial Number 42-221, September 25, 1942, *Correspondence Regarding the R-2160.*

[47] Col. F. O Carroll, Telegram from Materiel Division to Wright Aero, September 21, 1942 *Correspondence Regarding the R-2160.*

[48] Ernest C. Saltzman, R-2160 Development Report No. 58, Serial Number 42-221, September 25, 1942, *Correspondence Regarding the R-2160.*

[49] Shipping Ticket from Wright Field to W.A.C, October 8, 1942 *Correspondence Regarding the R-2160.*

[50] Opie Chenoweth, Memorandum Report Serial No. EXP-M-57-503-727, October 21, 1942 *Correspondence Regarding the R-2160.*

[51] Teletype from Engineering Division to Assistant Chief of Staff, February 9, 1943, *Case History of the R-2160*, - File.

[52] Col. H. Z. Bogert, Letter from Wright Field Technical Staff to Chief, Power Plant Laboratory, May 19, 1943, *Case History of the R-2160*, - File.

[53] Col. C. D. Gasser, Letter from Power Plant and Property Section to Brig. Gen. B. W. Chidlaw, Materiel Division, Washington, November 8, 1943, *Case History of the R-2160*, - File.

[54] T. M. Lucas, Letter from Wright Aeronautical Corporation director of Contract Administration to Commanding General, AAF Materiel Command, December 18, 1943, *Case History of the R-2160*, - File.

[55] Capt. Benjamin O. Kessel, Change Order No. 1 to Contract W535 AC-19438, February 12, 1944, *Case History of the R-2160*, - File.

[56] Product Chart, AAF Aircraft Equipment, 1945 (Aviation Hall of Fame and Museum Archives File 12-04).

[57] *Correspondence Regarding the R-2160* from August 31, 1930 through May 30, 1940.
See also Lt. H. M. McCoy, Memorandum Report EXP-M-52-592-5, February 9, 1940, *Case History of the R-2160* - File.

[58] *Case History of the R-2160*, – Brief.

[59] Arthur Nutt, Letter from Wright Aeronautical Corporation to the Office of Production Management, May 12, 1941, *Case History of the R-2160*, - File.

[60] *Correspondence Regarding the R-2160* beginning December 16, 1941.

[61] Brig. Gen. George C. Kenney, Assistant Chief, Materiel Division, July 23, 1941 Inter-office Memorandum to Chief, Materiel Division, Washington, D.C.

62 Lt. Col. B. E. Meyers, August 5, 1941 Inter-office Memorandum to Assistant Chief, Materiel Division, Wright Field.

63 Col. F. O. Carroll, Letter from Chief, Experimental Engineering Section to Wright Aeronautical Corporation, April 6, 1942, *Case History of the R-2160*, - File.

64 Col. F. O. Carroll, Inter-office Memorandum from Chief, Experimental Engineering Section to Chief, Power Plant Laboratory, May 23, 1942, *Case History of the R-2160*, - File.

65 Burnham Adams, Letter from Wright Aero to Commanding General, AAF Materiel Center, September 2, 1942, *Case History of the R-2160*, - File.

66 Raymond W. Young, Letter from Wright Aeronautical Corporation to AAF Materiel Center, September 3, 1942, *Correspondence Regarding the R-2160.*

67 Col. E. R. Page, Inter-office Memorandum from Chief, Power Plant Laboratory to Chief, Experimental Engineering Section, October 16, 1942, *Case History of the R-2160*, - File.

68 E. C. Sulzman, Wright Aeronautical Corporation Letter to Republic Aviation Corporation, September 19, 1942 *Correspondence Regarding the R-2160.*

6 Applications

The Republic XP-69, shown in mock-up above, was one of several aircraft designs that proposed use of the Tornado (Courtesy of Fairchild, via National Air and Space Museum, Smithsonian Institution, SI Neg No. 2001-823)

Aircraft designers seem to have viewed the Tornado with mixed feelings. While they participated in its initial definition, and saw it as a real solution to low-drag aircraft power without impossible structural and maintenance problems, few chose it initially for new designs. There were several reasons for this. First, the engine was only applicable to high-speed pursuit ships, attack aircraft, and medium bombers. Second, the X-1800 (H-2600) engine from Pratt & Whitney was more technically attractive. Third, the Pratt & Whitney X-1800 was thought to be much further along than the Tornado, and much further along than it really was. Still, the Tornado promised the greatest power of any available engine in 1939. Some designers picked other engines to power the first versions or their designs with the intention to retrofit the more powerful Tornado when it became available.

In a "competition" conducted by the Army Research Board in September and October of 1939, the large engines, either available or in development, were rated as follows:

Pratt & Whitney	X-1800	730.5 points
Wright Aeronautical	R-2160	715.5 points
Continental	H-2860	658.0 points
Allison	V-3420	610.0 points

In terms of development status, the Allison was already running, the Pratt & Whitney ready for test, the Wright in detailed design, and the Continental just a study [1]

Of the three engines that were actually built, specific weights were essentially equal:

Make	Model	Power	Weight	hp/in	hp/lb
P&W	X-1800	2,000	2,300[2]	0.77	0.87
Wright	R-2160	2,350	2,735[3]	1.09	0.86
Allison	V-3420	2,300	2,700[4]	0.67	0.85

For most aircraft designers, the first opportunity to utilize the Tornado came with "Request for Data R40-C," the fiscal year 1941 Army fighter procurement. A February 20, 1940, invitation to bid gave interested contractors ten days to request the R40-C data package. The contractors then had 30 days to submit preliminary design studies and engineering data. The Army reserved 60 days to evaluate the proposals and required the bids to remain open and valid for an additional 60 days after the evaluation period. Up to five designs were to be selected, and the contractor paid a maximum of $3000 for the winning designs. These five winners were then to participate in two additional program phases. Phase 2 authorized wind tunnel test models, stress analysis, and other engineering analysis needed to clarify a design. Phase 3 was for detailed design and construction of experimental aircraft to be tested by the Army.

Of the 26 R40-C proposals submitted, only three were to use the Tornado. These were the Curtiss P-249T, the McDonnell Model 1, and the Republic AP-12. The Curtiss P-249T was a Tornado powered variant of what was to become the XP-55 "Ascender." It was eliminated from the competition because it specified a contra-rotating propeller that was still in development. The McDonnell Model 1 was to become the XP-67 "Bat." It was eliminated from the competition because of low (if realistic) performance estimates and the proposed use of a complex system of shafts and gears to drive two pusher propellers from a central engine.

The Army evaluation board divided the proposals into three groups consisting of aircraft that were modifications of existing designs, departures from existing designs that had the potential for production in 1942, and designs that depended on further engine development. Three winners emerged, all from the second group. The Vultee Model 70, with a score of 817.9 (out of a possible 1,000), which was to become the XP-54 "Swoose Goose." The Curtiss P-249C, with a score of 770.6), was to become the XP-55 "Ascender." The Northrop N-2B, with a score of 725.8, was to become the XP-56 "Black Bullet." Two other designs, the McDonnell Model 1 and the Republic AP-12, with a score of 589.25, were selected for further study.[5]

In the final analysis, only four aircraft designs were serious candidates for the Tornado — the Hughes DX-2, the Lockheed XP-58, the Vultee XP-68 (a variant of the XP-54), and the Republic XP-69 (which is what the AP-12 became). In the following discussion, the DX-2 and XP-69 are covered in greater detail, largely because the other two have been subjects of numerous articles and author wishes to present previously unpublished information about the DX-2 and XP-69.

The Hughes DX-2

Drawing of the Hughes D-2A with Pratt & Whitney R-2800 Power. No photographs or drawings of the Tornado-powered DX-2 are known to exist. (National Air and Space Museum, NASM Videodisc No. 7A-27317, Smithsonian Institution)

Of all the aircraft designers to consider the Tornado, Howard Hughes showed the earliest interest and worked all the political angles necessary to obtain the engine for his high-speed fighter design, the DX-2. Despite Hughes' interest in the engine, details of the aircraft for which it was intended are extremely difficult to unearth. The project was shrouded in a level of secrecy that is probably unprecedented in the aircraft industry. Although photographs and drawings are thought to exist, none have ever been published.

Howard Hughes was obsessed with speed. On September 13, 1935 he set the world land plane speed record of 352.388 mph. The H-1B racing aircraft that was used to set this record had been built with two sets of wings. Hughes replaced the short wing with the long one and used the H-1B to set a transcontinental record of 7 hours 18 minutes 25 seconds from Burbank to Newark in January of 1937. Hughes then set a round-the-world record in July of 1938 using a Lockheed 14-N2. Some historians speculate that Hughes' desire for an even faster record breaker gave rise to the DX-2.[6]

Howard Hughes originally conceived the DX-2 airplane, whose later modifications were known as the DX-2A, D-2, XP-73, D-3, D-5, XA-37, XF-11, XR-11, and F-11. On December 5, 1939, the Hughes Aircraft Company expressed its desire to enter into a contract with the United States Army, agreeing, for the consideration of $50.00, to submit a report on the performance of a pursuit airplane under construction. After preliminary discussions, a contract, Purchase Order 40-07896, was signed May 22, 1940. This contract was for an engineering report of performance tests on this Hughes bombardment airplane and an engineering report on its suitability for mass production, setting forth changes in performance which would result from modifications necessary to meet Government military requirements. The Government was not obligated to buy the airplane.

Available information indicates that as of April 20, 1940 the DX-2 was a five-place, long-range medium bombardment airplane with a speed in excess of 300 mph and a normal gross weight of 37,000 pounds, capable of carrying a bomb load in excess of 4,000 pounds.[7]

At the request of Gen. Ira C. Eaker, then-Col. Oliver P. Echols on September 18, 1940 recommended authorization for Howard Hughes to buy two Wright Tornados for the DX-2. The procurement of these engines was to be negotiated directly with Wright Aeronautical. It was at that time impossible to predict delivery dates for the engines.[8]

By March 11, 1941, the design had been changed to a two-place fighter airplane. It had a wingspan of 60.5 feet, a length of 34.3 feet, a design gross weight of 31,000 pounds, a range of 2,600 miles, and a top speed of 450 mph. It was to be equipped with seven .50 caliber machine guns. The DX-2 was to be equipped with tricycle landing gear, and two Wright Tornado engines. Each engine achieved altitude performance through the use of dual parallel General Electric B-2 turbosuperchargers. The Tornados were to be fitted with huge 20-foot diameter, four-bladed Hamilton Standard propellers. Each propeller installation included a giant 40-inch diameter ducted spinner, with the spinner housing an axial-flow fan that fed cooling air to coolant radiators buried in the wings. Some additional information about the D-2 can be found in Appendix F—"Hughes D-2 Installation Engineering Details."

The airplane was to be made of plywood formed by the Duramold process, which made it particularly interesting to the Army, in view of the aluminum shortage. Wright Field recommended that all possible assistance in the way of technical advice, priorities, and necessary equipment be given to the project. Consequently, the Army Air Forces appointed Major Paul E. Shanahan as project officer and Allan Sullavan as project engineer to serve as liaison officers and the sole points of contact with the Hughes Aircraft. On September 4, 1941, an A-1-B priority was assigned to the development.

A test rig belonging to Howard Hughes was shipped to the Wright Aeronautical Corporation Flight Test Facility at Caldwell, New Jersey on July 15, 1941, and was expected to be operating with a Tornado installed by August 15, 1941. It was anticipated that this test rig would be very useful in determining overall Tornado heat rejection figures. In the Tornado, oil flowed at the rate of 300 pounds per minute. Oil entered the engine at a temperature of 180° F and left the engine at a temperature of 250° F. This was placing large demands on the oil cooler, whose size had already been determined. It was hoped that increasing inlet oil temperature to 200—210° F without changing the upper limit would decrease heat rejection through the oil cooler.[9]

David Rowe led the Hughes Aircraft team. What had been intended as a short summer test program grew into six months of nerve-wracking effort. The engine was not really fully engineered and presented a host of problems. An attempted start without a priming system led to a hydraulic lock in a lower cylinder and a buckled connecting rod. After much labor, about 30 hours was logged on the engine before a starting fire destroyed it.[10]

On October 1, 1941, representatives of the Hughes Aircraft visited Brig. Gen. O. P. Echols, Chief of Materiel Division, Washington, claiming difficulties due to lack of government interest in the project and requesting that the Government agree to buy the airplane if it should meet specifications. General Echols declined to make any commitments because of the secrecy with which the company had surrounded the development but promised to investigate the possibilities. A report from the Aircraft

Laboratory, Wright Field, on October 3, 1941 was extremely critical of the DX-2. The report advised that the combination of wood and metal planned for the wing construction would not work well together. The report went on to state that the Duramold construction would not withstand exposure to weather, that the airplane did not meet the Army Air Forces specifications for pursuit load factors, and that it was not equipped with standard military appointments such as armor plating.

On the basis of this and other investigations, Materiel Division informed Hughes Aircraft on November 7, 1941 that the DX-2 could not be made into a satisfactory military airplane and that the best interests of the Government would not be served by its procurement. Along with the letter, Materiel Division sent a general specification for a bomber escort airplane that would be acceptable to the government. [11]

However, this decision was reversed in a conference held on May 25, 1942, that was attended by General Echols, Brig. Gen. A. W. Vanaman, Commanding General of Materiel Division, Wright Field, and members of the Experimental Engineering Section of Washington and Wright Field. It was then decided that the DX-2 held sufficient promise of improvements in the aerodynamic and structural arts to warrant its purchase by the Army Air Forces. So, under the authority of Lt. Gen. H. H. Arnold, Commanding General of the Army Air Forces, Wright Field was directed to procure the DX-2. It was to be designated the XA-37, and no military equipment was to be installed on the prototype. Subsequent discussions resulted in an agreement between General Arnold and Mr. Hughes that the Government would absorb all of the development costs if the airplane proved sufficiently valuable after inspection and testing. By this time, Mr. Hughes had decided not to sell until after the DX-2 had flown, so its procurement was marked a "Hold Project". [12]

The change of mind over the DX-2 was probably a result of two factors. First, political advocacy came all the way from the White House. Elliot Roosevelt, a good friend of Hughes, was extremely interested in the aircraft for reconnaissance. Second, the Army Air Force had become impressed by the de Havilland Mosquito, which was also built of wood.

Back at Wright Aero, the 18-cylinder air-cooled R-3350 had become the highest priority engine project of World War II. Wright Aero had the dual problem of perfecting the design and getting the engine into mass production in time to start dropping bombs on the Japanese Mainland by the Spring of 1944. This had diverted nearly all Wright Aero engineering resources to the program.

Between February and April of 1942, it became clear that Tornado engines would not be available either in time or in sufficient quantity for all intended applications. The Army decided that R-2160s promised to Howard Hughes for the DX-2 would be diverted to the XP-58 project. Engineering Division made arrangements for Hughes to receive three Pratt & Whitney R-2800-49 engines instead.[13] The Pratt & Whitney powered DX-2 was to be known as the DX-2A.

Hughes at this time expected to eventually get Tornados for the DX-2 and continued engine testing with dual parallel General Electric Type B-2 turbochargers at Caldwell, New Jersey. In late 1942 or early 1943, General Electric witnessed some of these tests. At the request of the Army Air Force, General Electric made a number of recommendations for improving the turbocharger installation. These included:

- A means for inspecting, measuring and adjusting turbine wheel and cooling cap components
- Installation of a cooling cap of improved design
- Increased exhaust hood exit area and shape to prevent exhaust gas flow across the turbine wheel
- A "fishtail" extension to the waste gate to prevent turbulence from waste gate gasses entering the cooling hood
- A flexible joint in the exhaust stack before the nozzle box inlet
- Provision for a cooler and higher pressure cooling air source other than oil cooler exhaust air
- Improved flow of cooling air into and inside the turbocharger can
- Installation of thermocouples with quadruple shields rather than the unshielded variety installed on the test stand

Due to the secrecy that surrounded the engine installation on the DX-2, General Electric complained of being unable to obtain drawings and make any more detailed recommendations. In order to protect its interests in the turbocharger installation, General Electric requested that Materiel Center advise in regards to the proper approach to the situation.[14]

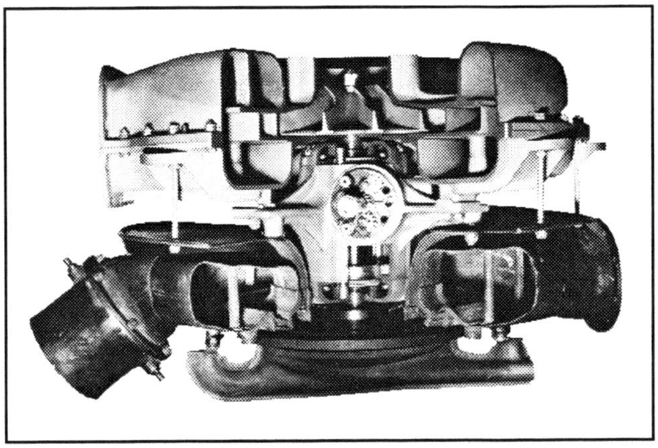

The ubiquitous General Electric Type B-2 supercharger was used in pairs on the Hughes DX-2. (National Archives and Records Administration)

Stanley A. Bell, Hughes Aircraft chief engineer, was appreciative of the "constructive criticism" from General Electric and went on to state that many of the recommended changes had already been incorporated on the R-2800 turbocharger installation on the DX-2A under construction at Culver City. However, since it was expected that a considerable time would elapse before the R-2160 engines would be installed, no changes to the test rig at Caldwell Field were contemplated.[15]

Using the R-2800s, the DX-2A (as it was now known) was flown by Howard Hughes on June 20, 1943 at Harper's Dry Lake in southern California. It underwent a long series of modifications to correct faulty lateral handling qualities, but was never really perfected. In the summer of 1944, on a clear day, the DX-2A was destroyed after the hangar in which it was stored was struck by lightening.[16]

Throughout the 1940s and into the 1950s, the Hughes Aircraft Company continued to revise the design and offer it to the government. The project ultimately grew into the F-11, a huge twin-engine reconnaissance airplane powered by Pratt & Whitney R-4360s. Howard Hughes was seriously injured in a crash of this aircraft on July 7, 1946.

The Hughes XF-11 was the ultimate evolution of the D-2. (Courtesy of the U.S. Air Force Museum)

Lockheed XP-58

The Lockheed XP-58 in its Final Form with Allison V-3420 Power (National Archives and Records Administration)

On March 11, 1939, the Army Materiel Division issued Circular Proposal 39-775 calling for a new type of twin-engine, high-performance interceptor fighter that derived as many design features as possible from existing aircraft.[17] Bellanca, Burnelli, Grumman, and Lockheed submitted proposals.[18]

The Lockheed Model 222 entry was a progressive development of the P-38 Lightning. It had the same general arrangement as the P-38, but featured a pressurized cabin and was to be powered by a pair of turbosupercharged 24 cylinder Pratt & Whitney X-1800-A2G (military designation XH-2600-5) liquid-cooled engines which were supposed to develop around 2,000 horsepower. Lockheed proposed to replace these engines by a pair of Wright R-2160 Tornado engines in production aircraft. Armament was to be a pair of 20-mm cannon and four 0.50-inch machine guns. Total fuel capacity was to be 300 U.S. gallons, as compared to 230 U.S. gallons for the early production P-38. The Model 222 was rather optimistically estimated to have a top speed of 473 mph at 20,000 feet when powered by the Pratt & Whitney XH-2600s, and a speed of no less than 500 mph at the same altitude when powered by the Wright Tornadoes.[19]

An Army review board met at Wright Field on August 3, 1939 to consider the four proposals that had been received. In a report issued on August 8, 1939, the board named the Lockheed Model 222 as the winner of the competition and recommended its procurement under the designation of XP-49. The second-place entry, Grumman Design 41, was a

The Grumman XP-50 (National Archives and Records Administration)

development of the Grumman XF5F-1 Skyrocket, a twin-engine carrier-based fighter. The review board recommended ordering one example of the Grumman entry as well. It was to be designated XP-50. A fixed-price contract, W535 AC-13476, for one XP-49 was entered into with Lockheed on January 8, 1940.[20]

In early 1940, the Army reserved the right to prohibit the export of U.S. made military aircraft to overseas customers. This was particularly true if the Army felt that the foreign order might result in delivery delays to its own squadrons. When Britain and France wanted to purchase the Model 222 Lightning from Lockheed, the Army was reluctant to approve the sale since it had already ordered YP-38 Lightnings for its own use. The Army finally did grant authorization to export the unturbosupercharged Lightning to Britain and France, under the condition that Lockheed agree to develop at no cost to the U.S. government a prototype of an advanced version of the Lightning, which was to be known as the XP-58. The formal agreement was signed on April 12, 1940.

This XP-58 was given the company designation of L-121. James Gerschler was project engineer. It was to be powered by a pair of turbosupercharged Continental IV-1430 liquid-cooled engines. Two versions were offered. The single-seat version was to retain the standard P-38 armament of one 20-mm cannon and four 0.50-inch machine guns. The two-seat version was to have an additional armament of a single 0.50-inch machine gun mounted in a remotely controlled turret at the end of each tail boom. Gross weight was estimated at 16,500 pounds, and the aircraft was expected to attain 450 mph at 25,000 feet.[21]

From the start, the XP-58 was a poster child for the procurement process gone wild. Since it was being developed using Lockheed money, the Army apparently felt free to order changes whenever it pleased. Lockheed had apparently gotten itself into a bind by its haste to obtain export approval for the P-38. Lockheed tried desperately to bring some order to the program, but the changes continued. While some of these changes were of Lockheed's own doing, there can be no doubt that had the program been better managed by the Army, the XP-58 may have been more than just a waste of money and manpower.

During a conference at Wright Field on May 6 and 7 of 1940, Lockheed suggested that the designations of the XP-49 and XP-58 be switched. The XP-58 was to be a baseline airplane, with the XP-49 evolving as a change order to the no-cost contract for the XP-58. In this manner, both aircraft could be built and differ only in number of crew, armament, and engines. Further, the Army would get 20 mm cannons, fuel and oil tank protection, a reduction in wing loading, an increase in range and armor plate protection for the XP-49 at no cost. The XP-58 development was covered under contract W535 AC-15867, approved on October 14, 1940.[22]

Design work progressed on the two aircraft. By the time the actual XP-49 change orders were written, it had been decided that the XP-49 would be overpowered with Pratt & Whitney H-2600 engines and that the XP-58 would be underpowered with Continental IV-1430s. It made sense to switch the powerplants between the two models. The change order was written to reflect this.[23]

On October 9, 1940, Lockheed's new project engineer, Neil Harrison[24], was informed that Pratt & Whitney had abandoned development of the H-2600. Engineering Section, Materiel Division suggested the Wright Tornado or Lycoming XH-2470 as alternative engines. Other engine recommendations were solicited from Lockheed as well.[25]

On November 4, 1940, Maj. F. O. Carroll, Chief, Experimental Engineering Section, was on hand for the first [public] run of the Wright R-2160 Tornado.[26] At that time, five engines were being constructed in Wright's experimental shops. Two were intended for loan to Howard Hughes and were to be ready for flight about July 1941; one was scheduled for the Power Plant Lab; the other two would remain at Wright for development testing.

Carroll was of the opinion that this was an insufficient number of engines. In order to get engines within one year additional orders would have to be placed soon and Carroll recommended purchase of five additional engines if Lockheed accepted the Tornado.[27]

Then, on November 13, 1940, Lockheed was informed that "The Materiel Division has decided to proceed with the installation of the Wright Tornado engine in this airplane. No provision will be made for the installation of any other engine. In view of the fact that the Wright company is proceeding rapidly with the development of this engine and also due to the delay which has existed in the construction of the XP-58 airplane, it seems now that this engine will be ready for flight at the time the airplane is completed."[28]

At this time, the XP-58 was about six months away from its first flight. In the author's view, it was profoundly naïve of Materiel Division to expect an engine with the complexity of the Tornado to go from first run to flight status in the same period of time.

In the mean time, Lockheed had been conducting its own engine investigation and had chosen the Pratt & Whitney R-2800 for installation in the XP-58. Despite the fact that a great deal of work had been wasted when the H-2600 was dropped, Lockheed had proceeded with work in other areas of the aircraft design. Lockheed felt that if the Army allowed them to proceed with the R-2800 installation, the project would be delayed by only two months. If, on the other hand, the Army insisted on the Wright Tornado, the XP-58 would have to be practically redesigned. With the heavier Tornado engines, balance considerations would force a change in the placement and plan form of the wing, as well as rearrangement of other equipment in the airplane. Further, Lockheed felt that in order to keep reasonable propeller diameter, dual coaxial propellers with two-speed drive would be required. Finally, with Tornado engines, the estimated gross weight had climbed to over 30,000 pounds.[29]

In the ensuing months, Lockheed attempted to extricate itself from its predicament, but to no avail. From the point of view of the Army, the Tornado powered XP-58 was a done deal. On November 27, 1940, procurement of three R-2160-3 (clockwise rotation) and two R-2150-5 (counterclockwise rotation) engines was initiated. One of each type was required for installation in the XP-58. Two additional engines of each type were required for spares. The third R-2160-3 engine was to be used for test in the Power Plant Laboratory.[30]

Lockheed calculated that use of the Tornado had resulted in an empty weight increase of 4,078 pounds, and submitted a claim for compensation in the amount of $82,350.[31] Lockheed now had grave misgivings about the airplane. With a wingspan of 70.0 feet, a length of 49.4 feet and a height of 13.6 feet, its gross weight was up to 33,785 pounds by February of 1941. That was two and one-half times the weight of the YP-38 upon which it was based and 8,000 pounds heavier than the Douglas DC-3! Lockheed was rightfully concerned about the maneuverability of such an airplane, and about the number of engineering hours that would be required to perfect it. While Lockheed admitted to a small share in the weight bloat of the project, most of the blame was laid squarely at the feet of the Government because of its shopping list of new requirements.[32]

Experimental Section was unwilling to change its assessment of the XP-58 priority. "... the XP-58, from a long range point of view, should be considered as a high priority development in view of the following:

> (1) More than a year has been wasted in getting any type of "accompanying fighter" under way.
>
> (2) The XP-58 is the prototype to prove and flight test the Tornado engine. (If we do not have this model, how are we going to fly the Tornado?)
>
> (3) It will be a demonstration article to determine the utility of a 33,000 lb airplane incorporating upper and lower rear turret protection for escort fighter purposes."[33]

Based on this input from Experimental Engineering, Lockheed was reminded of the importance of the XP-58, and that its continuation was warranted.[34]

The outside shell of an R-2160 was to be shipped to Lockheed in October of 1941 for installation in the XP-58. This shell engine was to be returned to Wright Aero, rebuilt, and returned to Lockheed by April 1, 1942.[35]

As of July 25, 1941, two engines were to be delivered to Lockheed on March 1, 1942 and April 11, 1942 for installation in the XP-58. [36]

At this time, eight months after the first run of the Tornado, only 44 hours had been accumulated on two prototypes.[37] It is no wonder that delivery schedules were starting to slip.

During the summer of 1941, while working with

Lockheed on XP-58 installation, Wright Aero developed a more efficient engine installation that reduced the installed weight of each engine by about 25 pounds at the expense of a 0.75 inch length increase. The length increase was not perceived to be significant.[38] Additional details on the engine installation appear in Appendix G—"Lockheed XP-58 Installation Engineering Details."

Lockheed established the date of August 25, 1941 for viewing of XP-49 and XP-58 mock-ups.[39] While there were a large number of changes, few affected the Tornado engine directly.[40] A change order was issued paying Lockheed $181,524 for additional costs associated with XP-58 development, including substitution of two Wright Tornado Model 689 engines for the original Pratt & Whitney X-1800-A2Gs.[41]

Correspondence during the first quarter of 1942 is scanty. It can be assumed that both the Army and Lockheed were fully absorbed in addressing the national emergency resulting from the bombing of Pearl Harbor, and that experimental projects took a back seat to production plans for existing war materiel.

By mid March of 1942, Lockheed was lobbying for the creation of a second XP-58 prototype. This was probably a result of experiences surrounding the crash of the XP-38 prototype at Wright Field on May 11, 1939.[42] This second airplane was intended to simply provide a backup in the event the first was lost. However, it was to go through a long and circuitous development path of its own. Delivery delays of the second pair of Wright Tornado engines contributed to this.[43]

From April through September of 1942, work on engine installation for the XP-58 was delayed because of fears that the Tornado dissipated too much heat into the cooling system. This ultimately turned out not to be true. A more detailed account appears in Chapter 5 – "The Tornado Program". On June 22, 1942, Wright Aero announced that delivery of the first XP-58 engine would be delayed until November. Still more personnel and testing facilities had been diverted to the R-3350 program.[44]

Meanwhile, back at Lockheed, the XP-58 engineering team had struggled through the dead time since the first engine delay back in April of 1942 This five-month delay presented serious problems to Lockheed because the aircraft engineering was essentially done, leaving the engineering staff with nothing to do. Consideration was given to conversion of the initial XP-58 into a low-altitude tank buster. This was attractive to Lockheed for an additional reason - wind tunnel tests indicated the XP-58 would be subject to the same compressibility problems that had plagued the P-38. A low altitude mission would allow less restrictive operation of the aircraft. In the final analysis, this was yet another change that went nowhere.[45]

On December 6, 1942, the propeller laboratory at Wright Field submitted a report evaluating the relative performance of the R-2160 in the XP-58. The results are summarized below.

> Single rotation four-blade propeller—Optimum full power operation was achieved wth a gear ratio 0.275:1. Optimum cruise operation was achieved with a gear ratio of 0.35:1 to 0.40:1. The single speed compromise was at a gear ratio of around 0.30:1
>
> Dual rotation six-blade propeller—Optimum full power operation (except take-off) was achieved with a gear ratio 0.30:1. Optimum take-off and cruise operation was achieved with a gear ratio of 0.375:1. The single speed compromise was at a gear ratio of around 0.325:1.

The additional weight (750 lbs. for dual rotation verses 575 lbs for single rotation) was not accounted for in this calculation.

While the dual rotation propeller was expected to yield increased performance over the single rotation propeller, the difference was marginal. Since the XP-58 allowed the use of a 14' 8" propeller, this was deemed adequate.[46]

In early February of 1943, Wright Aero informed Lockheed that delivery of the Tornado was delayed indefinitely. Lockheed was authorized to proceed with procurement and installation of the Allison V-3420s on February 12, 1943.[47] This was by no means the last change to the XP-58. Against all odds, with the fourth engine choice and innumerable other changes from its original concept, the XP-58 finally flew from Lockheed Air Terminal at Burbank on June 6, 1944. It was ferried to Muroc AAB where it made 25 flights that were marred by turbosupercharger problems. The XP-58 was delivered to Wright Field on October 22, 1944. Maintenance was such a problem that it rarely flew, eventually being used as a non-flying instructional airframe. Like so many other experimental projects that littered the ramp at Wright Field at the end of World War II, the XP-58 was probably scrapped. The second XP-58 was abandoned before it was ever completed.[48]

Vultee XP-68 (XP-54 variant)

Vultee XP-54 with Lycoming Power (Courtesy of the U.S. Air Force Museum)

The Vultee XP-68 "Tornado" was a proposed version of the XP-54 experimental pusher fighter that was to be powered by the Wright R-2160 Tornado radial engine driving a contrarotating propeller. The XP-54, or Vultee Model 70, was conceived as a fast interceptor using the Pratt & Whitney X-1800-4AG engine, and had been the winner of the R40-C competition. When Pratt & Whitney got out of the liquid cooled engine business, the aircraft was redesigned for the Lycoming XH-2470 and designated Vultee Model 84.

The XP-54 was a twin-boom, single-engine pusher with a span of 53 feet 10 inches. Its length was 54 feet 9 inches, and it was 13 feet high. It had a gross weight of 18,233 pounds and a useful load of just under 3,000 pounds. Its maximum speed was 381 mph at 28,500 feet. The maximum range was 500 miles. The XP-54 nose section had a very unusual design - it tilted slightly upward in order to increase the range of its two 37mm cannons. The dual .50-cal. machine guns were tilted slightly downward.[49]

The first mention of the Wright Tornado is in a request for quotation for second XP-54 with Tornado engines and one or two General Electric turbosuperchargers or two Wright turbosuperchargers and a pressurized cabin.[50]

In a telephone conversation between Maj. B. W. Chidlaw, Chief of Engineering Branch, and Mr. W. E. Burdick, vice president of Vultee Aircraft, Burdick informed Chidlaw that the XP-54 would have to be pretty much redesigned if it was to accommodate the Tornado. The Tornado added 600 pounds over the XH-2470, and required airframe changes. Further, Vultee faced a shortage of engineers if they were doing two parallel experimental developments. Finally, Vultee was looking for a way to fund the conversion of its prototype XP-54 into a production version.

Burdick wanted to know the status of the Tornado program. Chidlaw and General Echols thought (at that time) that the Tornado "presented a better picture than the XH-2470." Chidlaw expected an engine at Wright Field in August of 1941, but was candid that it was "particularly a test affair. You know the bugs that go into a '7 times 6' affair* like that". Chidlaw wanted to get the first article flying, and then go on to the Tornado installation.

Burdick still pressed the issue of redesign. Due to factors beyond the control of Vultee, weight had grown and pushed wing loading up into the 36 to 46 pounds per square foot range. With the Tornado, wing loading would be 50 or 51 pounds per square foot. Chidlaw asked about schedule and Burdick replied that the redesign to accommodate the Tornado would be completed around the middle of August 1941. Chidlaw said "To be frank with you, I don't think we have enough data on the Tornado to redesign for it intelligently." Burdick did, however, think Vultee had enough data to know that the high wing loading would make the aircraft low on service ceiling.[51]

Ultimately, Burdick's proposal was accepted. Vultee was directed to build the XP-54 to accept either the XH-2470 or the Wright Tornado. Both engines were to be turbosupercharged. The airframe was to be stressed to handle the 2,500 hp of the Tornado. Coolant radiators, oil coolers, and other accessories were also to be sized to handle a 2,500-hp powerplant. The first ship was to be Lycoming powered and was to fly in June of 1942. The second was to be Wright powered and anticipated delivery of the Tornado during July of 1942.[52] In July of 1941, the USAAF assigned the designation XP-68 to the project.[53]

During a conference on September 11, 1941, Vultee advised the Pursuit Project Office that the Tornado in its present configuration could not be used. The propeller was too large to fit between the twin booms. It was decided to discontinue work on the Tornado installation until it was available with contra-rotating two-speed gearing. One fact that emerged at this conference was that the XP-54

*Chidlaw is referring to the cylinder arrangement of the Tornado—seven rows of six cylinders each.

weight had grown from 15,000 pounds to over 18,000 pounds due to changes mandated by the Government. As a result, the aircraft had take-off distances that were entirely too long for fighter aircraft, a rate of climb that was too slow for an interceptor type, and insufficient fuel for one hour's flight at full throttle at 40,000 feet. In order to become an efficient military aircraft, the aircraft would have required Tornados with contra-rotating, two-speed props. But the Government was unable to supply these engines, so it was decided to build both prototypes using Lycoming power until the Tornados became real.[54]

The XP-68 was officially cancelled on November 22, 1941. The first of two XP-54s built made its initial flight on 15 January 1943 and flew 86 times before it was grounded by engine problems. The second aircraft flew only once.[55]

Republic XP-69

In 1940, Alexander Kartveli, vice-president and technical director of the Republic Aircraft Corporation, drew up plans for a new escort fighter using a single Wright R-2160 Tornado driving a contra-rotating propeller. Designated Model AP-18 by Republic, the aircraft was intended as a replacement for the P-47 "Thunderbolt". The engine was to be located immediately behind the pilot, and was to drive the propeller via an extension shaft to the nose. The cooling air intake and cooling system were located under the fuselage. The pressurized cockpit featured a large bubble canopy. A laminar flow wing was planned. Proposed armament was two 37-mm cannon and four 0.50-inch machine-guns, all mounted in the wings and firing outside the radius of the propeller.

Maximum speed was estimated to be 450 mph at 35,000 feet. The service ceiling was to be 48,900 feet, with a time to 35,000 feet of 20 minutes. Maximum range was to have been 1,800 miles. Estimated weights were 15,595 pounds empty, 18,655 pounds gross, and 26,164 pounds maximum. The Model AP-18 was to have had a wingspan of 51 feet 8 inches, a length 51 feet 6 inches, and a height 17 feet. The Army ordered two prototypes of the AP-18 in July 1941. The fighter was to be designated XP-69.[56]

Republic Aviation received a substantial amount of help with validation of the XP-69 design from the National Advisory Committee on Aeronautics

XP-69 Wind Tunnel Model (National Air and Space Museum, Smithsonian Institution, SI Neg. No. 2001-827)

Three-quarter-scale Mock-up of the Republic XP-69. Even this reduced-size version dwarfs the worker at the rear. (Courtesy of Fairchild, via National Air and Space Museum, Smithsonian Institution, SI Neg No. 2001-824)

A rear view of the XP-69 mock up shows details of cooling air exits. (Courtesy of Fairchild, via National Air and Space Museum, Smithsonian Institution, SI Neg No. 2001-823)

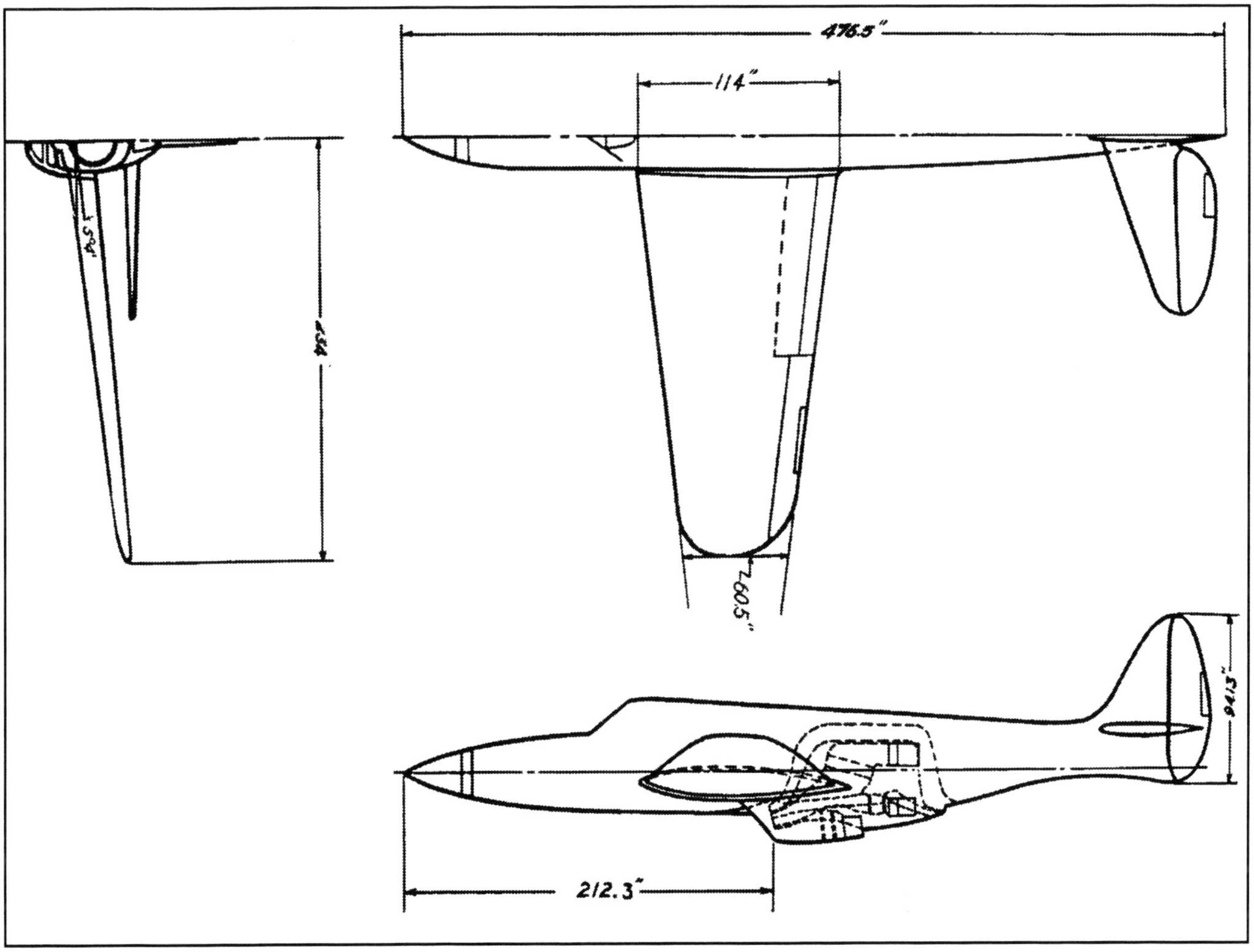

Three-view of the Proposed XP-69 (Courtesy of NASA Langley Research Center)

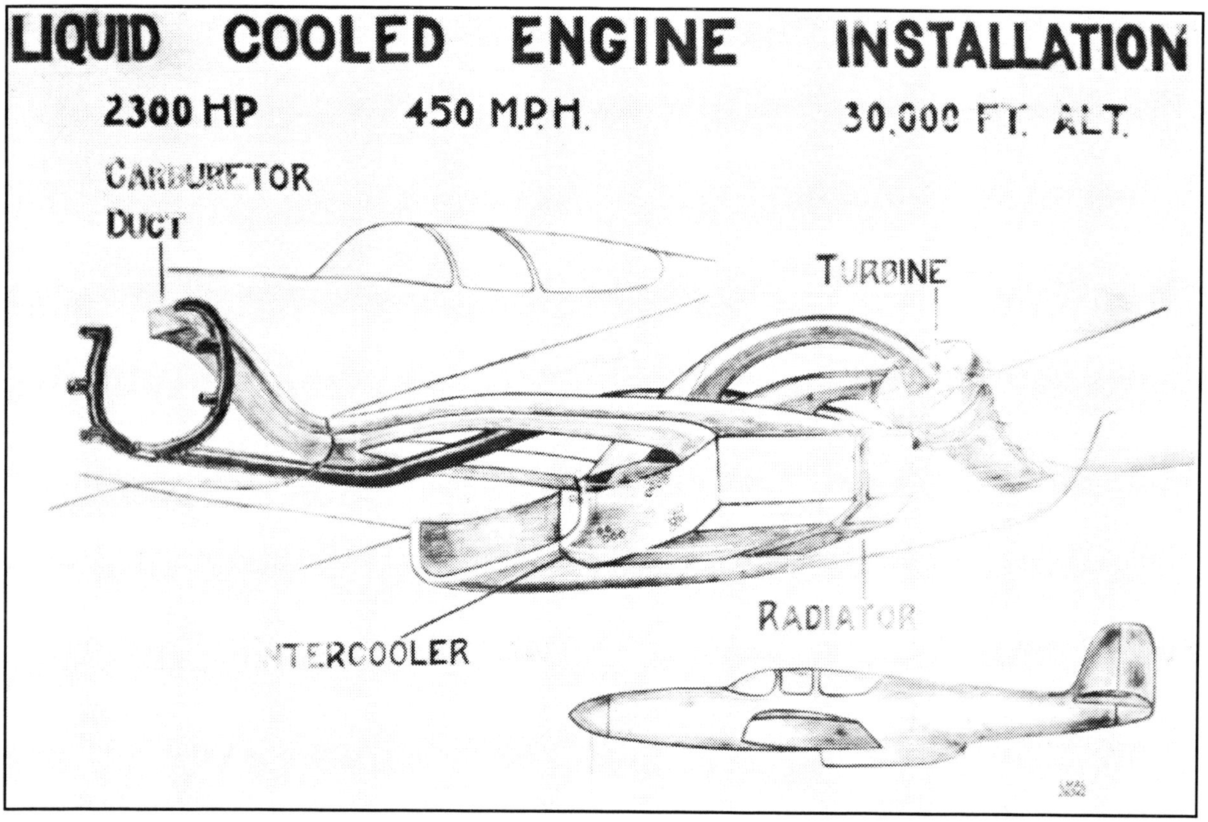

This early NACA study looks remarkably like a front-engine XP-69. The power rating, exhaust system, and carburetor air ducts are almost certainly for the Tornado. (National Archives and Records Administration)

(NACA) at Langley Field, Virginia. NACA was particularly useful in the radiator duct design, canopy shape, and compressibility effects over the wing and body. NACA had conducted a substantial amount of work on an Allison powered version of the "Thunderbolt" known as the XP-47, and had a large amount of data and a scale model that seemed applicable to the XP-69 program. When Republic approached NACA in August of 1941, two wind tunnel models were suggested. The first was to be used in the 19-foot tunnel for routine stability tests, and then cut down for compressibility tests in the 8-foot tunnel. The second full or large-scale model was destined to study cooling system and ducts in the full-scale tunnel. It was the modified XP-47 model that was considered for this series of tests. NACA also suggested live compressibility testing of an existing P-47B by installing a pair of total head tubes above the wing at a location between 60 and 80 percent of chord. Both were then connected to the airspeed pitot system through manometers. Evidence of shock would be given by a marked imbalance in the manometer readings.[57]

From the start of the XP-69 program, the aircraft was in competition and parallel development with another Republic design, the Pratt & Whitney R-4360 powered XP-72.[58] This approach allowed a head-to-head comparison of very similar aircraft with liquid-cooled and air-cooled powerplants. Gen. H. H. "Hap" Arnold preferred that Republic, with its experience as the Army's only builder of air-cooled pursuit airplanes, build the airplane with the R-4360. However, he was extremely anxious to get a Tornado engine installed in an airplane as soon as the engine was ready. To this end, General Arnold thought Curtiss Wright, with its history of liquid-cooled aircraft, should build its XP-62 with Tornado power instead of the R-3350. Curtiss was in the midst of several developments and had a shortage of engineering talent. Republic, on the other hand, had promised the XP-69 in fifteen months, and appeared to have the necessary available engineering to back this projection. General Arnold was willing to cancel the XP-62 if he could get reasonably early installation of the Tornado in some other aircraft.[59]

As negotiations progressed between Republic Aviation, the Army, and NACA, it became evident that most, if not all, of the stability data could be obtained from the model destined for the full-scale tunnel. In view of the backlog of projects bound for the 19-foot tunnel, this seemed the correct allocation of resources. While a model would be needed for the 8-foot high-speed tunnel as well, it could be a simplified one without empennage and with stub wings, somewhere between one-quarter and one-third full size. The full-scale model would need a functional propeller driven by an electric motor for two reasons: First, to simulate the slipstream effects in the stability, and second, to simulate the slipstream effects on airframe ducts.[60]

On September 29, 1941, the project was classified "Confidential", given the name "Development of the XP-69 Airplane", and assigned the classified designation MX-162.[61] Early tests were made on a one-twentieth-scale model of the XP-69 in the Langley 5-foot wind tunnel. The high speed of the XP-69 indicated by wind tunnel testing at 150 mph was predicted to be 440 mph with 2,500 hp at 35,000 feet. The landing speed was estimated to be 98 mph at sea level and gross weight. Longitudinal stability and control were satisfactory. Directional stability was more than normally required. Directional control was less than normally desired, even with the revised vertical tail. Lateral stability and control were considered satisfactory.[62]

During the fall of 1941, NACA was still busy assisting Republic with the XP-69 design, by supplying information related to the upcoming wind tunnel testing. Angle of zero lift, maximum coefficient of lift, and maximum angle of attack was provided for two airfoil sections, the NACA 66.2-313 and the 66.2-114. Drawings were also provided for installation of an electrically driven propeller in the model.[63] Republic also received the NACA report "Investigation of the Drag and Pressure Distributions over Windshields at High Speed."[64]

In the mean time, the Experimental Engineering Section at Wright Field had come up with a cooling system for the XP-69 that was believed, both by Engineering Section and NACA, to be superior to the Republic offering. Engineering section requested that provisions to include this revised cooling system in the three-quarter-scale model for tests in the full-scale tunnel. [65]

Apparently, there was interest at this stage of XP-69 development in the possible use of the Pratt & Whitney R-4360, or "X-Wasp" as it was known colloquially at the time. This notion was laid to rest in late December of 1941 when the Materiel Division, no doubt overwhelmed by the impending war effort, lost interest in the concept. NACA made repeated attempts to get clarification on the issue, but failed to get any response from Materiel Division.[66]

In a meeting between Mr. August Zinsser of Republic aircraft and the NACA Laboratory Staff, blueprints for the XP-69 model to be tested in the 8-foot high-speed wind tunnel were presented. It

XP-69 Particulars from Republic Model Specification No. 412 dated October 14, 1941.

General	
Weight, design gross	18,100 lbs
Power plant	WAC R-2160-3
Military rating	2500 hp at 4600 rpm at 35,000 ft
Normal rating	2300 hp at 4200 rpm sea level to 35,000 ft
Gear ratio, propeller	4:1
Propeller	two 3-blade, dual rotating
Propeller diameter	13.5 feet
Wing	
Wing area	508 sq ft
Span	52.0 feet
Mean Aerodynamic Chord	124.0 inches
Aspect Ratio	5.32
Incidence (relative to thrust)	1.75 degrees
Dihedral (L.E. wing)	6.0 degrees
Washout (Station 259)	2.5 degrees
Airfoil section(station 59)	NACA 66,2-114
Airfoil section (station 259)	NACA 66,2-213
Horizontal Tail Surface	
Total Area	102.00 sq ft
Stabilizer	65.20 sq ft
Elevator area, total	36.80 sq ft
Balance, percent of elevator area	22.0
Vertical Tail Surface, Original	
Total Area	49.50 sq ft
Fin, area	27.30 sq ft
Rudder, area	22.20 sq ft
Balance, percent of rudder area	16.0
Vertical Tail Surface, Area Reduced	
Total Area	38.50 sq ft
Fin, area	21.70 sq ft
Rudder, area	20.20 sq ft
Balance, percent of rudder area	20.2
Aileron Area, Total	50.20 sq ft
Flap Area, Total	66.60 sq ft

was decided that a one-fifth-scale model would be more suitable than the one-fourth-scale model previously contemplated. Mr. Zinsser also announced that a stress analysis would be sent by February 14, 1942.[67]

With a one-fifth-scale model of the XP-69 slated for completion by Republic on June 1, 1942, and a three-quarter-scale model to be completed on July 1, 1942, Materiel Division requested that NACA schedule two test series. The first was to be a compressibility study in the 8-foot high-speed wind tunnel to determine critical speeds of various components and modifications to increase these critical speeds if necessary. The second was to be a cooling and stability study of the three-quarter-scale model in the full-scale wind tunnel.[68]

On June 24, 1942, British Air Commission Group Captains H. W. Heslop and E. D. M. Hopkins reviewed the XP-69 Mock-Up and submitted comments. A report issued by the Army Air Forces revealed very few differences of opinion. The B.A.C. Report is quoted below, in full, with notes added on points disputed by the Army. *The Army notes are in italics.*

> A study of the exhaust pipe should be given to insure that a 'holed' exhaust pipe will not fire the fuel tank, which is close above the pipe. Further, it is important that the temperature of the fuel shall be kept low.
>
> The Glycol header tank is located to give a maximum frontal area for attack. Alternative location should be considered or some form of armour protection given.
>
> Pilot's Cockpit — The screen should be brought closer to pilot and provision made to enable the pilot to wipe the screen from inside – it is at present out of reach. As misting is a problem on aircraft at high altitude, some form of de-fogging should be considered both inside and out. *Army representatives do not concur with respect to the location of the windshield. De-misting and de-fogging of the windshield are being taken care of by heating.*
>
> Instruments — It is thought that as most pilots look over the left side to land, the flight instruments should be located on the left side and, further, as repeater compass is used it could be located on the instrument board.
>
> Map Case — Suggest a narrow case be placed forward on right side of cockpit near left leg, as well as existing case. *One map case will be supplied which is of a larger size instead of the two smaller ones.*
>
> Some indicator lamp should be placed to attract pilot's attention when cabin pressure fails, as additional to existing cabin altimeter. *The landing gear warning horn is to be used as a cabin pressure warning signal.*
>
> Trim Tab — These wheels should be made slightly larger, and while the elevator control is a crank, I personally prefer a wheel with indicator to show neutral position. *It is not believed that trim tab controls have to be larger due to the increased size of the airplane. The same controls are being used as on the XP-47B.*
>
> Armour — Some armour splash plate should be located at bottom of the armour glass screen. Also, consideration should be given to armour of cooling radiators. *Consideration has not been given to armor of cooling radiators due to the weight involved.*
>
> Undercarriage — A positive lock should be introduced on the control handle that prevents operation while the weight of the aircraft is on the chassis.
>
> Flap Control — This should indicate flap position if possible.
>
> Rounds Counter — Consider this desirable on aircraft of this size. *The size of the airplane should not dictate the military requirements for the rounds counter.*
>
> Sun Screens — Provision should be considered for sight, and also pilot's station against tropical sun.
>
> Navigation — Consideration of long range navigation will present serious problem.
>
> General
>
> A. It was thought that the armament installation was very well carried out, with great detailed consideration for quick turn round of the aircraft and loading of the ammunition boxes.
>
> B. The pilot's view was considered very good, apart from the points referred to in paragraph 3.
>
> C. The impression gained was that the aircraft was very large for a single engine fighter, and while appreciating the difficulties of incorporating all the special requirements for high altitude work, the question of maneuverability would have to be considered, bearing in mind the considerable weight which is placed aft,

which, the details, however, are not available for comparison with the P-47.

D. Navigation – The long range navigation would require some additional aids, such as the latest radio equipment. Radio equipment in accordance with present directives for airplanes of this type is installed in the aeroplane.[69 70]

On July 3, 1942, Hamilton Standard was alerted that an order was forthcoming for three propellers to be used on the XP-69. These were six-blade, dual rotation, automatic constant speed propellers having a diameter of 13' 8" and consisting of hub No. 625070 and blades 2C15A1-18 and 2C15A2-18.[71] A quote from Wright Aeronautical Corporation had been solicited on July 17, July 30, and September 8 of 1942, apparently producing no response.[72]

As of July 1942, the co-axial propeller drive for the XP-69 was still in the design stage. Tornado development was also proceeding slowly due to heavy demand on engineering resources by other engine programs, cylinder block cracks, connecting rod failures, valve failures, distribution difficulties, spark plug problems, and delays in ground testing the engine on the Hughes test rig.[73]

A series of brief progress reports from NACA issued from August 1942 through January 1943 details simultaneous work on several fronts. Each subject is discussed separately below.[74]

Cooling System

Controversy over the cooling system resulted in a number of letters being exchanged between NACA and Republic Aircraft. NACA, in a July 14, 1942 letter, worried of large pressure losses in the inlet and diffuser air ducting, and recommended that Republic have an alternate system in the event testing proved this to be the case. Republic's response on July 31 rejected NACA suggested alternative on grounds that it would require relocation of the upper longerons, installation of a preheat assembly in an already crowed equipment bay, would result in poor ram air pressure during climb, and would result in an intercooler frame of reduced strength. Gradually, Republic agreed that the duct system would have to be changed and sent two sheet metal workers to Langley on September 25, 1942. In a conference on October 2, 1942, Republic agreed to a complete redesign of the ducts, and to prepare new drawings as rapidly as practicable. The model was removed from the wind tunnel for over a month as

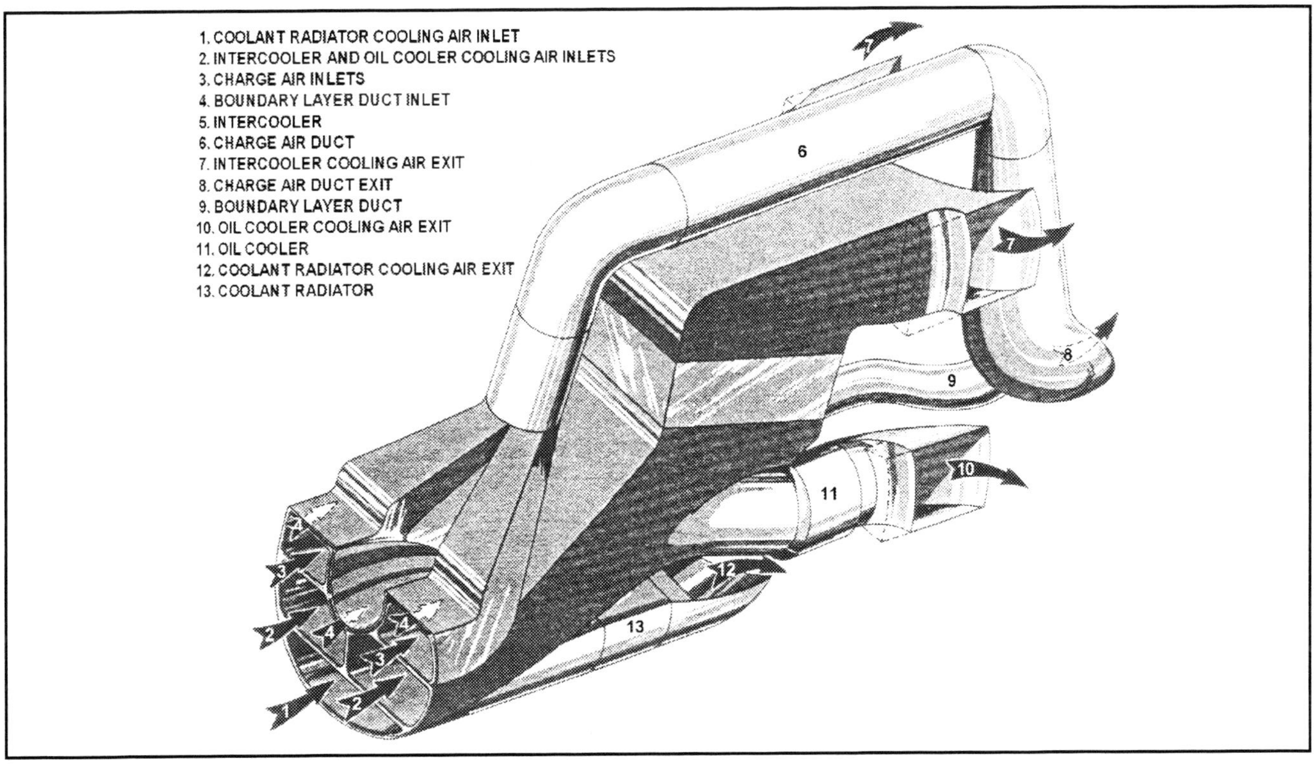

Duct Assembly Drawing for the Three-Quarter-Scale Model of the XP-69 Airplane (Courtesy of NASA Langley Research Center)

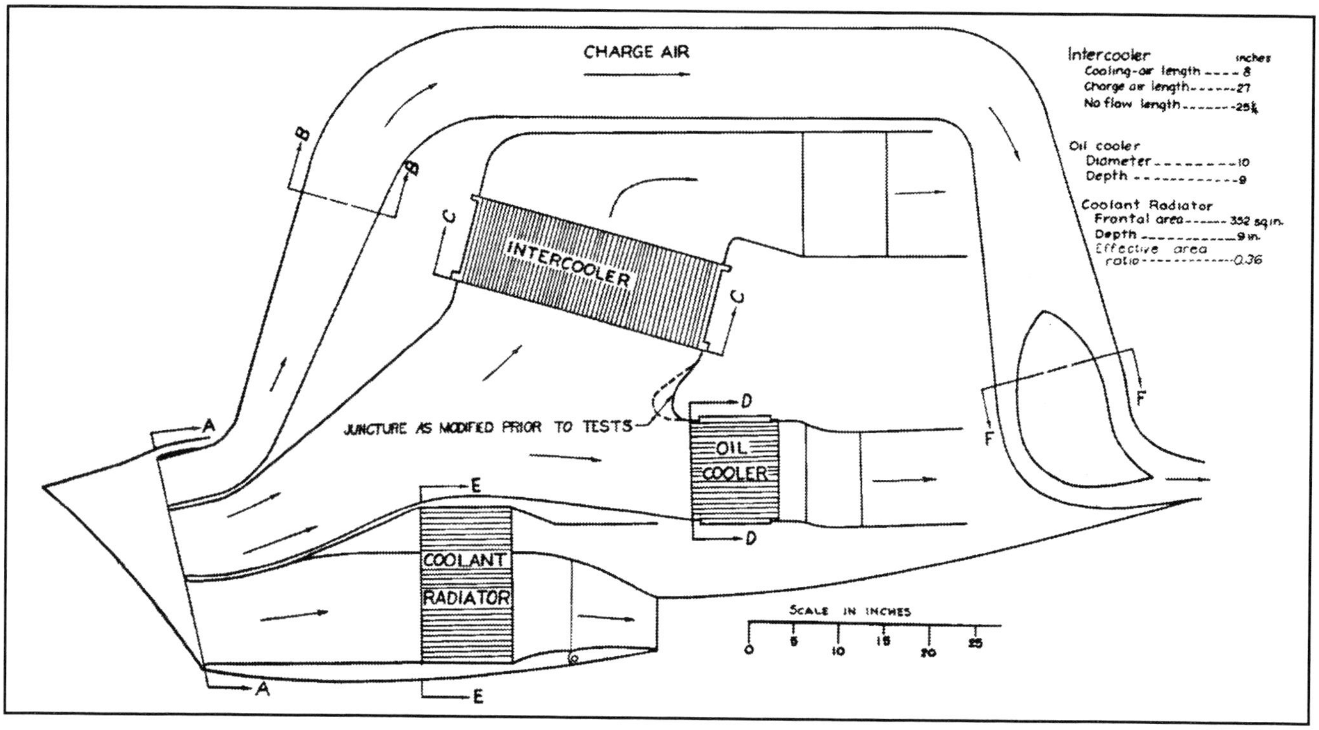

Schematic of the XP-69 Duct System (Courtesy of NASA Langley Research Center)

the duct system was reworked. Then, the model was returned to Republic for further work.

The Wright Field Aircraft Laboratory took advantage of the unsatisfactory performance of the XP-69 duct system and again pushed its own design that had first been announced back in October 1941.[75] However, in the 1st Indorsement supplied by Brig. Gen. F. O. Carroll, Chief, Engineering Division, the first hint of the final outcome of the XP-69 was telegraphed:

> "1. No action ... will be taken until the relative importance of the XP-69 project and the XP-72 project is established. If it is decided to push the XP-72 project, it will be necessary to hold the XP-69 in abeyance indefinitely, as facilities at Republic are not sufficient scope to handle both projects simultaneously.
>
> 2. If it is decided to let the XP-69 take preference over the XP-72, necessary action will be taken to substantially comply with your request. A decision on this matter should be made during the later part of January."[76]

NACA made several changes to the duct system, including extended boundary layer scoop, different

Front View of the XP-69 Scoop (Courtesy of NASA Langley Research Center)

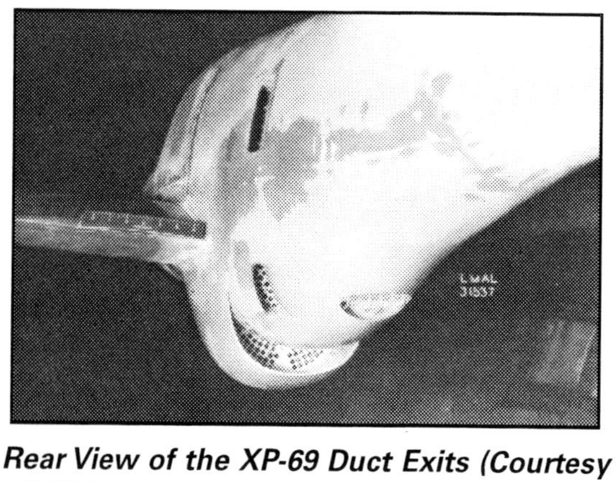

Rear View of the XP-69 Duct Exits (Courtesy of NASA Langley Research Center)

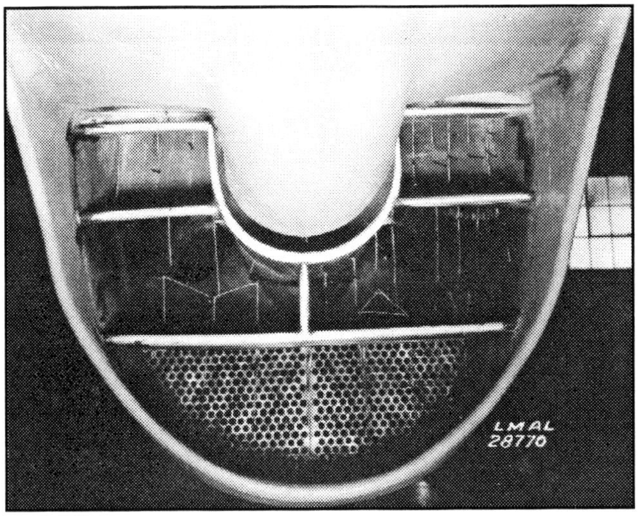

Close-up of Original Scoop (Courtesy of NASA Langley Research Center)

scoop shapes, and internal vanes in the intercooler and oil cooler ducts. None of these produced any significant improvement in pressure recovery.[77] It is not known whether the Wright Field duct design was ever tested.

High Speed Study

The one-fifth-scale model for the 8-foot high-speed wind tunnel, initially promised for February 14, had been delayed until August 18 because of manufacturing difficulties resulting from extremely damp weather. August 18 arrived, but instead of a model, NACA received a telegram from Republic

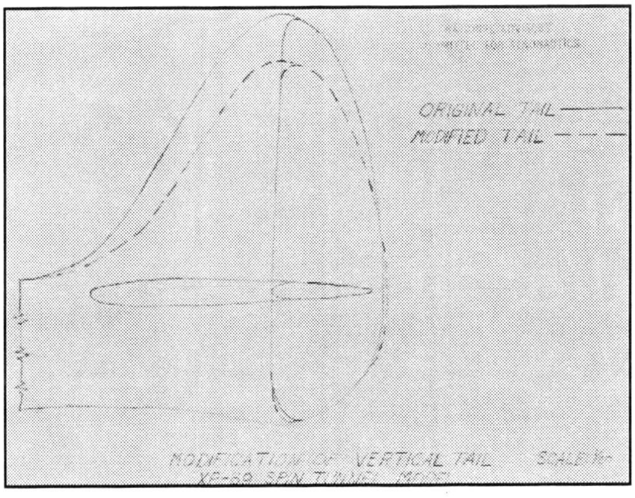

The XP-69 tail was altered as a result of spin tests. (National Archives and Records Administration)

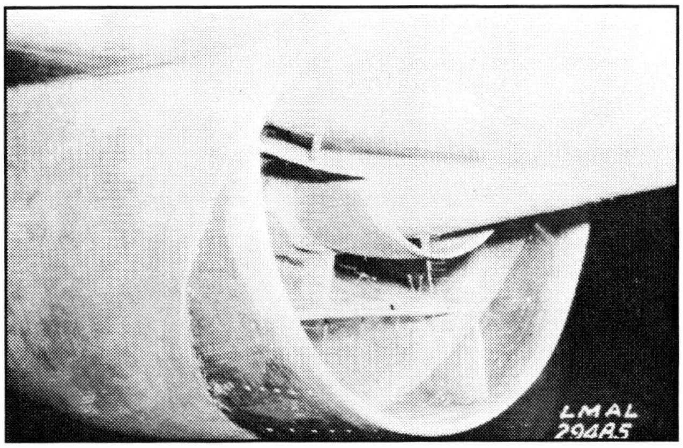

Close-up of the Revised Scoop (Courtesy of NASA Langley Research Center)

announcing a ten-day delay, again because of damp weather. Some work was done on a model of the tail in a two-dimensional, two-foot tunnel, but proceeded slowly at first as the test team refined its technique. The model for the 8-foot tunnel finally arrived on December 22, 1942

Spin Tests

On August 6, 1942, NACA issued its report on XP-69 spin tests, finding the spin characteristics generally satisfactory. As a result, Republic Aircraft reduced the size of the vertical tail and resubmitted the spin model for additional testing which were for all practical purposes, as good as the first model.

Full Scale Tunnel Tests

One interesting finding of the full-scale tunnel testing, which had begun on August 25, 1942, was that the aileron forces were too high. Several solutions were proposed, including an internal sealed balance with wedge trailing edge. Use of a balance tab was also discussed. Previous work on the P-47 indicated that little more could be done by increasing the amount of balance. The P-47 ailerons were about ninety percent balanced. Due to its larger size, the XP-69 would require an even higher degree of balance. One point in favor of the XP-69 was the fact it was fitted with a control wheel permitting about 135° of travel from neutral versus the stick in the P-47. This provided approximately twice the mechanical advantage of the P-47 control stick. Some wind tunnel data indicated that the proposed wedge trailing edges would probably overbalance the ailerons for some conditions. Moreover, a straight trailing edge in place of the cusp trailing edge that was part of the airfoil section would be almost 100 percent balanced. This was considered to be too delicate a situation in view of aileron and wing distortion resulting from maneuvering loads. NACA recommended that Republic consider some sort of booster rather than trying to balance the ailerons too closely.[78]

An Engineering Division Memorandum complained about the 14° angle between the main landing gear strut center line and vertical, pointing out that the *Handbook of Instruction for Airplane Designers* dictated an angle of five degrees or less. It was recommended that the *Handbook* requirement be adhered to.[79]

An Abrupt End to the Project

Despite Republic's progress on the fighter, the Tornado engine project was at a standstill. Brig. Gen. F. O. Carrol's prediction of the past December came true: "Higher authority decreed, by letter dated May 11, 1943, that subject contract with the Republic Aviation Corporation for two (2) XP-69 airplanes would be terminated. It is therefore mandatory that all possible be salvaged from equipment being procured for installation in these airplanes."[80] Republic's parallel XP-72 project showed more promise because of the choice of the Pratt & Whitney R-4360 powerplant.

While the XP-69 had been an entirely new development, the XP-72 was evolution of the P-47. Two XP-72 prototypes, Army serial numbers 43-36598 and 43-36599, were ordered on June 18, 1943. XP-72 Serial No. 43-36598 flew for the first time on February 2, 1944. The second prototype flew for the first time on June 26, 1944.Impressed by the performance of the prototypes, the Army ordered 100 aircraft. However, this order was cancelled before production started.[81]

The Republic XP-72 (National Archives and Records Administration)

Summary

It should be clear by now that designing airplanes around unproven engines is risky business. The same scenario as outlined above, that of good airplane designs failing because of an unfinished engine, has been repeated dozens of times. The only example of a relatively successful airplane that entered service during World War II with an immature engine is the Boeing B-29. The cost of achieving this, in terms of engineering effort, dollars, and human life, was staggering. Some would argue that without a need, engine makers would never improve their product. Still, there are precious few examples of successful aircraft built around experimental engines.

---

1 Schlaifer, *The Development of Aircraft Engines*, 288-289.

2 *X-1800-A4G Engine Specification No. A- 7019.*

3 *Wright Tornado Engine Specification No. 617-C.*

4 Whitney, *Vee's For Victory*, 286.

5 Balzer, "Request for Data R40-C."

6 Baugher, *P-73.*

7 *Case History of the Hughes D-2,* frame 1071.

8 Routing and Record Sheet, Chief of Army to Chief, Materiel Division, August 30, 1940, *Case History of the R-2160,* – File.

9 W. W. Woodward and O. F. Carlson, Memorandum Report EXP-M-57-503-390, July 2, 1941, *Case History of the R-2160,* – File.

10 Boyne, "The Most Elusive Hughes," 50.

11 *Case History of the Hughes D-2,* frames 1071-1072.

12 Ibid, frame 1072.

13 Col. B. W. Chidlaw, Memorandum from Experimental Engineering Section, Wright Field to Gen. Echols, April 21, 1942, *Case History of the R-2160,* – File.

14 E. S. Thompson, Letter to Commanding General, AAF Center, Wright Field, January 6, 1943, *Correspondence Regarding the R-2160.*

15 Stanley A. Bell, Letters to AAF Materiel Center, Wright Field, February 10, 1043 and March 17, 1943, *Correspondence Regarding the R-2160.*

16 Boyne, "The Most Elusive Hughes," 54.

17 *Case History of the XP-58*, frame 1467.

18 Letter from the Circular Proposal 39-775 Review Board to the Chief of the Army, August 8, 1939, *Case History of the XP-58*, frame 1504.

19 Baugher, *P-49.*

20 *Case History of the XP-58*, frame 1479.

21 Baugher, *P-58.*

22 *Case History of the XP-58*, frames 1480 - 1481.

23 Change Order No. 1 to contract W-535 AC-13476, August 21, 1940, *Case History of the XP-58*, frames 1530 - 1537.

24 Baugher, *P-58.*

[25] F. O. Carroll, Chief, Experimental Engineering Section in a letter to Lockheed Aircraft, October 9, 1940, *Case History of the XP-58*, frames 1539-1540.

[26] F. O. Carroll, Major, Inter-office Memorandum from Chief, Experimental Section to Chief, Power Plant Laboratory, Wright Field, November 6, 1940, *Case History of the R-2160,* – File.

[27] Ibid.

[28] F. O. Carroll, Letter to Lockheed Aircraft, November 16, 1940, *Case History of the XP-58*, frame 1549.

[29] Hall L. Hibbard, Lockheed Vice President and Chief Engineer, Letter to Assistant Chief, Materiel Division, November 19, 1940, *Case History of the XP-58*, frame 1551.

[30] *Case History of the R-2160,* – File.

[31] Hall L. Hibbard, Letter to Assistant Chief, Materiel Division, December 26, 1940, *Case History of the XP-58*, frames 1558 – 1560.

[32] Robert E. Gross, President of Lockheed, Personal Letter to O. P. Echols, Chief, Materiel Division, February 10, 1941, *Case History of the XP-58*, frames 1561 – 1564.

[33] Major B. W. Chidlaw, Chief, Experimental Engineering Branch, Inter-office memorandum to General O. P. Echols, March 18, 1941, *Case History of the XP-58*, frames 1566 – 1568.

[34] General O. P. Echols, Letter to Robert E. Gross, President of Lockheed, April 4, 1941, *Case History of the XP-58*, frame 1574.

[35] W. W. Woodward and O. F. Carlson, Memorandum Report EXP-M-57-503-390, July 2, 1941, *Case History of the R-2160,* – File.

[36] Routing and Record Sheet, July 14, 1941, *Case History of the R-2160,* – File.

[37] Ibid.

[38] T. M. Lucas, Letter from Wright Aeronautical Corporation to Assistant Chief, Materiel Division, August 5, 1941, *Case History of the R-2160,* – File.

[39] F. O. Carrol, Telegram to Experimental Engineering Branch, August 6, 1941, *Case History of the XP-58,* frame 1575.

[40] F. C. Wolfe, Memorandum Report on Mock-up Inspection of the XP-58 Airplane, Sept 15, 1941, *Case History of the XP-58*, frames 1578 – 1586.

[41] Change No. 1 to Contract No. W535 AC-15867, September 30, 1941, *Case History of the XP-58*, frames 1587 – 1593.

[42] Hall L. Hibbard, Letter to Assistant Chief, Materiel Division, March 16, 1942, *Case History of the XP-58*, frame 1595.

[43] Col. A. E. Jones, Chief, Contract Section, Letter to Lockheed Aircraft, May 29, 1942, *Case History of the XP-58*, frame 1602.

[44] Maj. M. C. Draler, Memorandum Report to Colonel Chidlaw, June 24, 1942, *Case History of the R-2160,* – File.

[45] Hall L. Hibbard, Letter to Commanding General, Materiel Center, September 3, 1942, *Case History of the XP-58*, frame 1604.

[46] Col. H. H. Couch, Memorandum from Chief, Propeller Laboratory to Chief, Engineering Division, December 26, 1942, *Case History of the R-2160,* – File.

[47] Brig. General A. E Jones, Teletype to Lockheed Aircraft, February 12, 1943, *Case History of the XP-58*, frame 1634.

[48] Baugher, *P-58.*

[49] Ibid.

[50] Lt. Col. A. E. Jones, Telegram to Vultee Aircraft, May 28, 1941, *Case History of the XP-54*, frame 1156.

[51] Major B. W Chidlaw, Transcript of telephone conversation with Mr. Burdick of Vultee, July 7, 1941, *Case History of the XP-54*, frames 1159 – 1163.

[52] Major J. F. Phillips, Approval of recommendation of Wright Field Committee by General Echols, July 16, 1941, *Case History of the XP-54*, frame 1164.

[53] Baugher, *P-54.*

[54] Maj. M. S. Roth, Report on Conference with Vultee, September 19, 1941, *Case History of the XP-54,* frames 1166 - 1168.

[55] Baugher, *P-54.*

[56] Baugher, *P-69.*

[57] Col. Carl F. Greene, Materiel Division Liaison Officer, Letter to Assistant Chief or Materiel Division, Wright Field, August 29, 1941, *Correspondence Regarding the XP-69.*

[58] Lt. Col. F. O. Carroll, Letter to Materiel Division Liaison Officer at NACA, September 5, 1941, *Correspondence Regarding the XP-69.*

[59] Gen. O. P. Echols, Telegram to Assistant Chief, Materiel Division and reply, September 29, 1941, *Correspondence Regarding the XP-69.*

[60] Elton W. Miller, NACA Head Mechanical Engineer, Memorandum for Engineer in Charge, September 18, 1941 *Correspondence Regarding the XP-69.*

[61] Lt. Col. F. O. Carroll, 1st Indorsement to Chief, Technical Staff, Wright Field, September 29, 1941, *Correspondence Regarding the XP-69.*

[62] *Test of 1/20 Scale Model of Republic XP-69.*

[63] Col. Carl F. Greene, Materiel Division Liaison Officer at NACA, Letter to Republic Aircraft, October 30, 1941, *Correspondence Regarding the XP-69*

[64] Col. Carl F. Greene, Materiel Division Liaison Officer at NACA, Letter to Republic Aircraft, November 10, 1941, *Correspondence Regarding the XP-69.*

[65] Col. Carl F. Greene, Materiel Division Liaison Officer at NACA Laboratory, Letter to Assistant Chief, Materiel Division, Wright Field, October 25, 1941, *Correspondence Regarding the XP-69.*

[66] Col. Carl F. Greene, Materiel Division Liaison Officer at NACA Laboratory, Letters to Assistant Chief, Materiel Division, Wright Field, December 17, 1941, January 17, 1942, and February 9, 1942, *Correspondence Regarding the XP-69.*

[67] H. A. Fedziuk, NACA Assistant Aeronautical Engineer, Memorandum for Engineer in Charge, Langley Field, February 3, 1942, *Correspondence Regarding the XP-69.*

[68] Col. T. A. Sims, Assistant Technical Executive, Materiel Division, Letter to Dr. G. W. Lewis, Director, NACA, Washington, April 25, 1942, *Correspondence Regarding the XP-69.*

[69] Group Captain H. W. Heslop, British Air Commission, Letter to Col. B. W. Chidlaw, Chief Experimental Section, Materiel Command, Washington, June 24, 1942, *Correspondence Regarding the XP-69.*

[70] Col. F. O. Carroll, Chief, Experimental Engineering Section, Wright Field, Letter to Assistant Chief of Staff, Materiel Center, July 13, 1942, *Correspondence Regarding the XP-69.*

[71] Col. F. O. Carroll, Chief, Experimental Engineering Section, Wright Field, Letter to Hamilton Standard Propellers, July 3, 1942, *Correspondence Regarding the XP-69.*

[72] Col. A. B. Jones, Letter to Wright Aeronautical Corporation, September 8, 1942, *Correspondence Regarding the XP-69.*

73 Maj. N. C. Gemler, Memorandum Report on Visit to Wright Aeronautical Corporation, July 24, 1942, *Case History of the R-2160,* – File.

74 Col. Carl F. Greene, Materiel Division Liaison Officer at NACA Laboratory, Progress Reports to Commanding General, Materiel Division, Wright Field, August 21, 1942, September 14, 1942, October 22, 1942, December 3, 1942, January 11, 1943, *Correspondence Regarding the XP-69.*

75 Col. Paul H. Kemmer, Chief, Aircraft Laboratory, Memorandum to Chief, Experimental Engineering Section, Wright Field, November 27, 1942, *Correspondence Regarding the XP-69.*

76 Brig. Gen. F. O. Carroll, Chief, Engineering Division, 1st Indorsement to Kemmer Memorandum, December 15, 1942, *Correspondence Regarding the XP-69.*

77 Nelson, *Investigation of the Duct Systems.*

78 Floyd L. Thompson, NACA Principal Aeronautical Engineer, Memorandum to Chief of NACA Aerodynamics Division, March 29, 1943, *Correspondence Regarding the XP-69.*

79 Col. Paul H. Kemmer, Chief, Aircraft Laboratory, Memorandum to Chief, Engineering Division, Wright Field, April 19, 1943, *Correspondence Regarding the XP-69.*

80 Col. H. Z. Bogent, Chief, Technical Staff, Engineering Division, Letter to Chief, Power Plant Laboratory, May 19, 1943 *Case History of the R-2160,* – File.

81 Baugher, *P-72.*

7 Restoration

Years of abuse and neglect are apparent in this 1988 photograph of the Tornado (Courtesy of the New Jersey Aviation Hall of Fame and Museum)

Obscurity

The only remaining Tornado was delivered to Wright Field in 1945. Little was known about it until it appeared in the back lot of the New England Air Museum (NEAM) at Bradley Field, Connecticut. At some point, the Tornado, along with a huge collection of other engines and equipment was transferred from Wright Field to the newly formed Air Force Museum at Wright Paterson Air Force Base, Ohio.

Like many of the experimental projects from World War II, the Tornado faded into such profound obscurity that its existence was nearly forgotten. It left the Air Force Museum as scrap sometime in the 1960s. According to Harvey Lippencott, former NEAM director of exhibits, Air Force personnel in charge of the Air Force Museum had several house cleanings and threw out many of the experimental test engines to make room for the continuous arrival of new artifacts. In those days, political appointees ran the show, and believed the old engines were of questionable value and had no historical significance.[1]

Correspondence from the 1970s between individuals and the National Air and Space Museum (NASM) document at least a passing interest in the engine.[2] Ironically, one of those interested was Col. Richard Uppstrom, then director of the U.S. Air Force Museum, the very institution that had sold the engine for scrap a decade earlier! Initially, NASM could locate only a few photographs. Gradually, as warehouses of material was processed, additional information was found. A few specifications and reports, including technical descriptions, longitudinal and cross section drawings, and photographs were discovered.

Sometime after the engine was scrapped, a reclusive Ohio collector of aircraft parts and books named Walter Soplata purchased the engine from a Cleveland scrap dealer. Soplata apparently held the engine for a time. Then in the late 1970s, he arranged for a collection of artifacts, of which the Tornado was an incidental part, to be donated to the New England Air Museum.[3] It was received there late in 1980, and inspected by Harvey Lippencott on December 1, 1980. According to Lippencott, many of the donated items had come directly from the disposal yard at Wright Paterson Air Force Base in the 1960s. Air Force Museum tags identified these. The Tornado had no such tag.[4] The engine sat under a tarp in the back lot of the museum for nearly a decade.

For people interested in engines, NEAM is a wonderful place, a literal treasure trove of history. Besides NEAM's large and impressive exhibit collection, hundreds of engines and airplane parts are scattered over many acres of land in the back lot. Much of the credit for this collection goes to Harvey Lippencott.

Harvey Lippencott

Lippencott grew up in Moorestown, New Jersey where he developed an early interest in aviation. A Quaker, he served as a technical representative for Pratt & Whitney in Europe during World War II instead of fighting. After the war, he completed his engineering degree at Georgia Institute of Technology. He then returned to Pratt & Whitney and worked as an engineer until the early 1970s, when he became the first corporate archivist for its parent company, United Technologies. After retiring in 1987, Lippencott became the volunteer archivist and exhibit director for NEAM.

Lippencott's association with NEAM went back to its founding meeting in 1959. Lippencott used his pull with United Technologies, along with his vast network of industry and museum contacts to expand the NEAM collection at every opportunity. In addition to his work at NEAM, Lippencott served as president of the American Aviation Historical Society. As a result of his work in aviation history Lippencott gained notoriety as a leading expert. Shortly before the new National Air and Space Museum opened in 1976, Lippencott spent five days reading every single label to make sure everything was correct. Lippencott died on December 30, 1996.[5]

Jerry Abbamont

According to all that participated in the acquisition and restoration of the Tornado, it is Gerard A. "Jerry" Abbamont whose interest, vision, and persistence are responsible for saving the engine. Jerry was born October 5, 1918 in New York City. As Jerry neared that impressionable age of ten, the age when the dreams of the child begin to shape adult aspirations, his life was changed forever. Like so many other boys of that era whose contributions to aviation are inestimable, Jerry waited spellbound during late May of 1927 as Charles Lindbergh flew solo from New York to Paris. For Jerry, that single day began a lifelong love for all things associated with aviation. He attended Central Building and Trade School in New York City from which he graduated in 1935. While in school, he worked at Sky Harbor Seaplane Base, Hudson River, New York City as a mechanic. He also worked at Marine Aviation Space at North Beach which was to eventually become La Guardia Airport. There, each evening, it was his duty to light the kerosene lamps that served as runway lights.

Gerard A. Abbamont— "Foremost C-W Historian" (Courtesy of Edith Abbamont)

In the mid-1930s, Jerry worked on Ford Tri-Motors that were used for barnstorming. In April 1940, he joined Wright Aero where he was continuously employed until 1983. During World War II, Jerry worked as a tester, foreman, and general foreman. It was during this time that he became acquainted with the Tornado.

In 1945, Jerry moved with Wright Aero from Paterson to Wood-Ridge, New Jersey, where he was a manager for about fifteen years, involved primarily in experimental engine assembly and testing. By 1960, Wright Aero was starting to fall on hard times. In the 1950s, Wright Aero enjoyed considerable success with the R-3350 engine in commercial airline applications. The company management literally thought this would go on forever. But Curtiss-Wright, which had been the second richest company in the world at the end of World War II, was unwilling to commit the necessary corporate funds for research and development.

The reciprocating market disappeared in the early sixties, and Curtiss-Wright never successfully competed in the gas turbine market. By the late sixties, Curtiss-Wright had become a second-tier job shop with its aviation roots rapidly dying. The company tried a laundry list of other markets, including nuclear power for submarines. Jerry participated in this operation from 1969 through 1976. Curtiss-Wright then got into gas turbines for electrical power generation. These were originally fueled by natural gas, and later by coal using pressurized fluidized bed combustion.

Jerry managed the assembly and test of generator sets and the pressurized fluid bed pilot project until his retirement in 1983. In every case, the upper

Part of the "Year of Aviation in Paterson" Exhibit (Courtesy of Edith Abbamont)

management of Curtiss-Wright, which since the beginning had been composed of investment bankers, were unwilling to spend the money for research and development that would have kept Curtiss-Wright going. It was this same problem that had caused Frederick Rentschler to leave Wright Aeronautical in 1925 to form Pratt & Whitney Aircraft. In nearly sixty years, little had changed.

Despite the short sightedness of Curtiss-Wright management, Jerry loved the company and its people. He was smart, likeable, and remarkably well connected. He stayed in touch with the people he had worked with, and saw some of them from time to time at gatherings of retired workers.

After his retirement, Jerry took his grandson to the Paterson Museum to see if there was an exhibit on the history of Curtiss-Wright. The Paterson Museum is headquartered in an old industrial complex near a waterfall that had brought pre-Civil War silk and cotton mills to the city. This formed the nucleus of an industrial base that eventually hosted the largest manufacturing center for steam locomotives in the United States. As steam traction engines became obsolete, Wright kept the town alive. It seemed only natural that the Paterson Museum would have some sort of tribute to Curtiss-Wright. But there was none. By 1986, Jerry had assembled a Curtiss-Wright exhibit for the museum, using artifacts scavenged from the scrap yard at Wright Aero.

To Jerry, the Curtiss-Wright part of the *Year of Aviation in Paterson* exhibit was permanent, but the Paterson Museum had other ideas. In 1987, Jerry got a call informing him that space was needed and asking him to please come and get the exhibit he had so painstakingly assembled. Having no place to put it, he began checking around the area and found the New Jersey Aviation Hall of Fame and Museum (AHoF) in Teterboro. AHoF is dedicated to preservation of New Jersey contributions to aviation.

As the museum had little information on Curtiss-Wright, Jerry was welcomed with open arms. AHoF had no engines in its collections. That soon changed as director Pat Reilly was smitten by Jerry's infectious enthusiasm. The AHoF engine collection and the Curtiss-Wright Archive at the AHoF became a full time hobby for Jerry. As he later related in a letter to another retired Curtis-Wright employee, "I recognized that if I had it to do all over again, AHoF would have been the original location." He was soon putting in 35 hours a week at the museum, and he and his wife Edith worked at home constantly.

Jerry was determined to preserve the legacy of Wright Aeronautical. He single-handedly assembled the largest collection of Wright Aero material in the world and found it a permanent home at AHoF. Much of this was quite literally salvaged from company dumpsters as the investment bankers cleaned house. Other items Jerry collected from the donations and private collections of ex-Wright Aero employees.

Those who knew Jerry and were familiar with his love for aviation and boundless energy expected it to go on forever. That is why it came as such a surprise when in 1991 Jerry came down with a sickness that the doctors couldn't diagnose and that would not respond to treatment. His health deteriorated to a terrible state. He could no longer write, and had to depend on Edith to prepare his correspondence. In one of his last acts, he made sure that Graham White had plenty of information for his book *Allied Aircraft Piston Engines of World War II.* White remembers that despite being confined to bed and in an extremely weak state, he insisted on talking for hours in an effort to preserve as much as possible of the Wright Aero history. Jerry died quietly at home on June 3, 1994.

Jerry's friends still speak of him with great admiration. He was a natural leader whose excitement was contagious. He met his wife, Edith, because of his participation in a volunteer service that provided transportation to music lessons for blind children. He worked with a number of organizations for retired Wright Aero personnel. He was instrumental in raising money for AHoF. Mostly, he is remembered as warm and quiet, always the gentleman. "Jerry never got nasty with people, his family or employees. He always had a very even temper and was seldom angry," remembers Edith. "When we were married, everyone told me I had married the greatest guy in the world. He was the most wonderful man I have ever known."

Mel DeBlock, one of the Tornado restoration team, recalls "Jerry was a good man, a family man. From the beginning, he let people know that he didn't do things halfway. He was very thorough. You knew that if you were involved in a project with him that he expected you to be involved as well, and carry the load you had signed up for."

Emerson Watson, another of the Tornado team recollects, "Jerry was the spark plug of that museum. He knew a lot of people and was very smart and very persuasive. If it weren't for him, they wouldn't have that Tornado engine."

Dave Way, another restorer, reminisces, "Jerry was a go-getter. It is entirely due to him that AHoF got the engine. Harvey Lippencott was key. He and Jerry got along famously. Jerry convinced him it was unlikely NEAM would get around to restoring the engine. If NEAM would lend it to AHoF, they would restore it. By the time five years passed, no one at NEAM wanted it back. They were more interested in restoring airplanes, and that is what their restoration people were involved with. It was an ingenious sales campaign that Jerry put on."

Augie Zoll took over the Curtiss-Wright Archive at Jerry's request. Augie says of Jerry, "His memory was so great and he talked to so many people over all the years. He could talk all day and never repeat himself. The guy was amazing."

William T. "Bill" Figart, then Vice-President of the Rotary Engine Division of John Deere Technologies, Inc., credits Jerry's tireless work in preserving the history of Curtiss-Wright. At one of the Curtiss-Wright retirees' reunions, Figart presented Jerry with a beautiful plaque. On it was a chromed half-inch thick slice from the center of an R-3350 piston and an inscription reading "Gerard A. Abbamont — Foremost C-W Historian". Figart did not realize how much this memento meant to Jerry until he learned that Jerry had requested it be buried with him.[6]

The Dream of Bringing the Tornado Home

Of all the engines and projects Jerry had been involved with, the Tornado was his favorite. He had personally tested both the 14-cylinder prototype and the complete engine during his early days at Wright Aero, and had always hoped to locate one of the engines. All traces had long since disappeared from Wright Aero. During one of the work sessions at AHoF, the crew got around to remembering the Tornado and dreaming about how great it would be to locate one and bring it back to New Jersey. Jerry had tried unsuccessfully to locate a Tornado for years. He found the engine listed as part of the NEAM collection in the 1982 edition of *Aircraft Engines in Museums Around the World,* published by NASM. NEAM advised him that the engine had been disposed of.

One of the AHoF restorers, while visiting NEAM on unrelated business, spotted an engine he thought might be the Tornado. During the summer of 1986, Jerry and Edith traveled to NEAM to see if there really was a Tornado. He was skeptical because another Tornado, a V-12 built by Wright in the 1920s, had long caused confusion. Jerry found an odd engine under a tarp, and, upon closer inspection, identified it as the later Tornado. He turned to Edith and grinning said, "I'll be back later!"[7]

And go back he did. Of all the words that characterize Jerry, the most frequently used is "persistent." Jerry returned to AHoF and began building the political framework necessary to give the Tornado a permanent home. He first enlisted the help of AHoF executive director Pat Reilly, and then began to court influential members of the AHoF board of directors.

In September of 1987, AHoF Executive Director Pat Reilly sent a letter to NEAM curator Robert North suggesting the possibility of swapping the Tornado at NEAM for two J65 jet engines at AHoF. Reilly knew NEAM was restoring a Martin bomber and thought engines were needed to complete that restoration.[8] North's response was brief and to the point. The B-57 already had engines and was nearly completed. He further stated, "We do have a Wright 'Tornado' but at this time have no desire to part with it. We are just getting our engine exhibit organized and this engine has a definite place in it. We do, however, appreciate your interest in it and would hope to let you know if at any time it becomes surplus to our needs."[9]

Undaunted by this initial rebuff, Jerry traveled to NEAM in the spring of 1988 to pitch the idea of getting the engine on loan and restoring it at the Teterboro School of Aeronautics. This was formalized in a letter from AHoF trustee David G. MacMillan to NEAM executive director Mike Speciale.[10] AHoF correspondence files do not show how this proposal was received. Since it took nearly two more years for AHoF to get the engine, it must be assumed that the proposal was not well received.

Still, Jerry persisted. Over the weekend of October 7-8, 1988, he attended a meeting of the Northeast Aero Historians. Here he met Robert B. Meyer, Jr., former engine curator for the NASM, and Harvey Lippencott. A friendship grew between Jerry and Lippencott that played the pivotal role in securing the loan of the Tornado.

Upon his return to Teterboro, Jerry sent Lippencott a collection of souvenirs given to Curtiss-Wright employees by the company and a photo of the "Silver Dart," a 1909 aircraft built by the Aerial Experiment Association with whom Glenn Curtiss had been associated. With this donation was a letter outlining Jerry's extensive aviation experience, his intense interest in the Tornado, and his desire to somehow get it restored.

As he became better acquainted with Lippencott, Jerry felt confident in asking for favorable recommendations of a proposal presented to the Committee of Exhibits at NEAM. This proposal was simple:

> AHOF&M members (principally retiree's [sic] and former C.W. employees), many of whom had worked on the engine, in its design, manufacture, assembly and test would assist in its removal to AHOF&M in New Jersey for refurbishment to display quality, within the scope of what could be reasonably accomplished. No effort would be made to dis-assemble same but more likely would be a restoration that we have done to other engines in our collection. That would consist of removal of all external components, such as Magnetos, ignition harness, spark plugs, Carburetor, and fuel lines, Coolant pumps and lines. Each of these items would be painted in the original color scheme as well as the engine housings, cylinder banks, etc. The transport truck would be cleaned, serviced, and painted, hopefully the castors could be re-used. The engine would be fixed in the flight position, a Curtiss Electric Propeller, cut down to keep its size within the envelope, installed. This propeller would be operational in pitch change only and would have a console with the necessary controls. The completed unit which would take about a year to accomplish would be moveable within a complex and transportable over the road via a flat bed type car carrier.
>
> The costs associated with this refurbishment to be borne by AHOF&M of N. J. We at AHOF&M are in a position to offer several items that are in excess to our collection as well as continued assistance in the future for information that we have on C.W. history.

Harvey Lippencott traveled to AHoF on March 23, 1989 to examine the collections and capabilities of the museum and its restoration team. This meeting was another key in the eventual transfer of the Tornado. In a brief March 25 letter to Jerry, Lippencott wrote, "I think we all arrived at a better understanding and appreciation of each other and what we are trying to do. I also obtained a better understanding of Pat and the HF Museum program. I encourage you to keep at the Wright Aero project. It needs to be done. You have my support. At the next curatorial meeting (April 15) I will put the R-2160 on the docket. As I explained to you and Pat, there is no chance that NEAM will release ownership. A restoration/loan consideration may fly. I now have a much better understanding of what your group and the HF can do and I have a warmer feeling toward your proposal. Your proposal has much merit."

In his reply to this letter, Jerry formally requested a loan of the Tornado for a period of five years, and outlined the plans he had for it. He anticipated that restoration would take one and one-half years. For the remainder of the time, the engine would be a central part of activities observing the 50th anniversary of Curtiss-Wright participation in World War II. Jerry reiterated that if the engine ever became surplus that AHoF desired first priority for its acquisition.

At the April 5, 1989 meeting of the NEAM curatorial committee meeting, Lippencott presented Jerry's proposal and gave it his endorsement. The real issue was whether the five-year loan of the Tornado would impact the planned experimental propulsion exhibit at NEAM. Lippencott made the points that:

> The restoration workmanship at AHoF was of excellent quality, as good as that of NEAM.
>
> There were advantages to having the restoration done by people familiar with Wright Aero practices and processes.
>
> The NEAM experimental propulsion exhibit was of such magnitude that outside restorers were going to have to be used anyway.

Examination of the engine while still at NEAM showed minor damage and nearly 45 years of dirt and grime. Note the dented valve cover. (Courtesy of the New Jersey Aviation Hall of Fame and Museum)

Preparation of the experimental propulsion exhibit would take about five years, so the Tornado would return at a good time. Besides, space could be made for it if the exhibit was completed early.

The NEAM curatorial committee accepted the proposal based on a signature of a formal restoration and loan agreement. Lippencott was to draw up restoration and preservation guidelines, and were to make periodic inspections as the work progressed. Harvey Lippencott died before the NEAM Experimental Propulsion Exhibit could be completed. If it had been carried out, it would have been the largest of its kind in the world.

Preparations for the Return of the Tornado

On May 22, 1989, Jerry informed Lippencott of a change in the restoration work site. Originally, AHoF had proposed doing the work at Teterboro School of Aeronautics. In the interim, it had become evident that it was better to do this work in the homes of the restorers than at the School of Aeronautics. Not only was it more convenient, but it was also significantly more secure. Access to the school was available a maximum of thirty hours per week, with no access on weekends or holidays. Also, there had been losses of tools and parts. For these reasons, Jerry proposed restoring the Tornado in the garage of Ken Bell, a restorer who had worked with Jerry for decades, and about whom we will hear more later. Apparently there was no objection from Lippencott, because this is exactly what happened.

As soon as Jerry knew for sure that the Tornado was coming to AHoF, he began to exploit his extensive network of former Curtiss-Wright employees. He produced a display outlining the project for the 1989 Wright Aero Jamboree held in Cocoa Beach, Florida. Since Jerry was unable to attend the meeting himself, he asked Bill Figart to take the

Physical damage and corrosion to the right coolant pump and associated plumbing required many hours of reconstruction. (Courtesy of the New Jersey Aviation Hall of Fame and Museum)

history with him. Figart was a former Curtiss-Wright vice president who moved on to John Deere Technologies when rights to the Curtiss-Wright rotary combustion (Wankel) engine were sold. Included with the display was a plea for help, which produced a number of responses, all of which were relayed to Jerry through Figart.

One response came from John McElroy, who had been a field engineer for Wright Aeronautical and had worked on both the Lockheed XP-58 and Hughes DX-2 projects. McElroy had two Tornado coolant pumps, which were needed to replace the badly corroded ones on the actual engine. Since he knew nothing about AHoF and since AHoF did not yet have the engine, McElroy was squeamish about parting with the pumps. Jerry sent him a long letter and some photographs, telling the story of how the engine had been located, how NEAM had acquired the engine, and how AHoF had convinced NEAM to loan it the engine. In less than a month, Jerry had the coolant pumps and a collection of other photographs, videotapes, and technical documentation as well.

One sticky issue that surfaced repeatedly during negotiations for the engine was the degree to which it would be torn down. Initially, NEAM had wanted a restoration to the same specifications used by the NASM. In this plan, the engine would have been completely torn down, and every part refurbished. Any deviations to this plan would have been discussed on a case-by-case basis with NEAM. Aside from the prohibitive costs and schedule delays, the lack of special tools and technical data would have almost surely resulted in doing more harm than good to the engine.

Faced with the prospect of a restoration effort that was significantly more complicated and costly that had first been envisioned, Jerry traveled with Ken Bell, Bruce Scott, and Emerson Watson to NEAM on August 26, 1989 for a detailed inventory of the Tornado's condition. Scott was a former chemical specialist who had worked in materials engineering at Curtis-Wright. Watson was formerly of the Curtis-Wright flight test section. All agreed that for an engine that had not run for more than 45 years, it was in pretty good shape. For AHoF purposes, cleaning and painting would result in a very presentable display. To restore it to NASM standards, while not complicated, would be costly. Everyone agreed to proceed with the restoration and to work out the details as the work progressed. The really big item, identified by Scott, was that all of the exposed fasteners would need to be re-plated with cadmium. By 1989, the United States Environmental Protection Agency had targeted plating shops nationwide as large producers of heavy metals that found their way into ground water. Cadmium was one of the worst offenders. Because of the crackdown, plating shops were closing by the hundred, and those staying in business were raising rates to astronomical levels to pay for the increased pollution controls.

Two things happened to solve the impasse. First, NEAM volunteered to get the parts plated at a facility it regularly did business with, one with more reasonable rates than those polled by AHoF. NEAM even agreed to foot some of the restoration costs. Secondly, NEAM relaxed the requirements for restoration to NASM standards. Still required was a boroscope inspection of internal engine components and disassembly to the point that such an inspection could be made. Otherwise, the restoration could proceed with minor disassembly, and whatever cleaning, preserving, and repainting necessary to produce a museum quality exterior finish.

With the final impediments pushed aside, both parties signed the official loan agreement on November 25, 1989.

The Restoration Team

While negotiations for the engine progressed, Jerry had also been busy assembling a restoration team. These were Kenny Bell, Mel DeBlock, Emerson Watson, Dave Way, and Augie Zoll. [11]

Kenny Bell performed the majority of Tornado restoration tasks. (Courtesy of Mel DeBlock)

Kenny Bell

Like Jerry, Kenneth "Kenny" Bell had actually worked on the Tornado. Kenny was born in Atlantic Highlands, New Jersey on August 1, 1918. From May 23, 1944 until January 20, 1946, he served in the U.S. Navy as a Motor Mechanic Mate aboard the amphibious cargo ship USS Union.

Kenny and his wife Ingrid never had children. They enjoyed travelling, in the both U.S. and abroad. Kenny restored old cars and participated in some car shows. He was also active in his community. He built a gazebo in a small park across from his home and kept the bus stop benches in the neighborhood repaired and painted.

It is not known when Kenny came to Wright Aero. According to Dave Way, "Kenny was a very competent test foreman. Had respect of the guys that worked for him as well as the engineering staff that worked with him. He was very organized and very inventive in terms of solving problems. When Curtis-Wright abandoned the reciprocating engine business, Kenny got involved with the rotating combustion engine and was in charge of assembly and test of that engine."

Jerry and Kenny had known one another since Jerry arrived at Wright Aero. When Jerry began work on the exhibit for the Paterson Museum, it was Kenny who Jerry called on to help. Together, they assembled the exhibit, doing whatever restoration work was required on all of the engines that were displayed. Kenny was also in on the ground floor of work at AHoF and specifically the Tornado restoration. Since the restoration was done in Kenny's garage, he worked on the engine nearly every day for the one year and three months it was there. During that time, he kept a detailed log that chronicles the huge amount of time that went into the restoration. The log also gives insight into Kenny's perfectionist philosophy. It was this approach to his work that produced the beautiful finished product that the Tornado now is. For Kenny, the restoration was a labor of love, and accomplished despite health problems and other hardships. Kenny suffered from emphysema and had difficulty breathing. Much of the time, Kenny worked at the end of a 30-foot oxygen hose. He died on November 1, 1997 at the age of 79.

Mel DeBlock

Mel DeBlock was born in Lodi, New Jersey on April 25, 1922. He grew up in Lodi and graduated from Lodi High School in 1939. He did a bit of private flying at Teterboro airport, and enrolled in Casey Jones School of Aeronautics in Newark at night. This background provided an entrée to the third shift Experimental Assembly position at Wright Aeronautical starting January 16, 1942.

Mel DeBlock (left) did much of the beautiful lock wiring on the restored Tornado. This photograph was taken in Kenny Bell's (right) garage after completion of the restoration. (Courtesy of Mel DeBlock)

Mel enlisted in the Army Air Force after World War II started but was rejected because of his eyes. In April of 1942, he went back and passed the physical. It was October 19, 1942 before he actually entered service. He was trained in a gunnery program, ultimately becoming an instructor, and finally a Gunnery Instructor instructor. Near the end of the war he served in Guam and Saipan.

After the war, Mel returned to Curtiss-Wright where he was a union steward for 13 years. When Mel decided to give up that obligation, Jerry offered him a supervisory position in Experimental Assembly. When development of new engines ceased, Mel was transferred to the Overhaul Department where he was responsible for teardown and cleaning of the J-57. In 1969, he transferred to Power Systems Division where he coordinated engineering, vendor, and assembly activities. After Power Systems ceased operation, Mel supervised the packing, shipping, and receiving until the plant closed in November of 1983.

While Mel did not personally work on the Tornado during his early years in Experimental Assembly, he remembers seeing it from time to time. Experimental Assembly was a very restricted area. The Tornado was assembled in a room at the end of the department that was even more restricted. Only three or four persons per shift were allowed access, and Mel was not among them. However, by seeing the engine being moved in and out of its room on its way to the test stand and by talking to people with access, a bit could be gleaned about the engine.

Emerson Watson (left) hides behind a Tornado ignition harness in Kenny Bell's garage. On the right is Bruce Scott, a lubricants and coatings consultant. (Courtesy of the New Jersey Aviation Hall of Fame and Museum)

When it became necessary to move the engine exhibit from the Paterson Museum to AHoF, Jerry asked Mel for help. When Jerry found the Tornado, Mel came to work on the project. Mel was one of those who lobbied to avoid complete teardown because of the lack of assembly equipment and technical data. When the restoration team had problems removing the rear accessory cover, it was Mel who finally sorted out the hidden fasteners and arcane disassembly procedure that had turned the job into a huge puzzle. Finally, it was Mel, with his years of experience in Experimental Assembly, who got the duty of replacing the hundreds of feet of safety wire on the exposed fasteners of the valve covers and other exterior engine parts. This highly visible contribution of Mel's is a true work of art.

Em Watson

Emerson Watosn was born in Atlantic City, New Jersey on October 8, 1918. He attended Atlantic City High School and graduated in 1935. He always has been and is still interested in aviation. As a kid, he was an avid model maker. Em went to work for Wright Aero as a Junior Mechanic on May 6, 1941 at their Experimental Flight Test Section at Caldwell Airport in Caldwell, New Jersey. There, he earned his Airframe and Powerplant Mechanic's license. He progressed to Mechanic, Senior Mechanic, Crew Chief, Aircraft Maintenance Foreman, General Foreman and finally Superintendent of Experimental Flight Test, Hangar Operations. Flight Test Operations moved from Caldwell to Edwards Air Force Base, California in 1953. The Edwards operations closed and Em moved back to Caldwell in 1961 to work with the VTOL Division of Curtiss-Wright. Em recalls, "The 26 years with Experimental Flight Testing were the best and most enjoyable of my life." He was in charge of modifying and maintaining 13 aircraft Curtiss-Wright used for various projects. These included such strange birds as a B-17 with a Constellation nacelle in place of the original nose. It was to test the R-3350 Turbo Compound and was equipped with something highly unusual for an airplane – mufflers!. There was also a B-29 with a jet engine in the bomb bay, and the Curtiss-Wright X-19 Vertical Take-Off and Landing aircraft. After the X-19 crashed in 1965 on its 50th flight, the program was cancelled. With the cancellation of the last of Curtiss-Wright flight test operations, the Caldwell facility was closed and Em was offered a position as a foreman in Experimental Assembly back at the Wood-Ridge, New Jersey plant. Here he remained until the experimental assembly section was closed in 1966. After years of being outside on the flight line, the plant seemed like a dungeon. One bright spot was that his boss was Jerry Abbamont. Though he rarely saw Jerry, they gradually became acquainted. When Jerry started his restoration project for the Paterson Museum, he called on Em to help.

Em restored a number of engines now on display at AHoF, including the 2-cylinder Lawrence and the Reaction Motors XLR-25 engine used in the Bell X-2. As of this writing, he is working on restoring a Reaction Motors XLR-11, used on early X-15 flights.

Em's recollection of the XLR-25 restoration shows how Jerry got things done. "The XLR-25 was just a bunch of pieces. We had no idea what it was supposed to look like. By getting pictures and prints from the Smithsonian, we were able to figure out how it went together. Parts came from different places. One guy had the exhaust nozzle in his garage. Another had the turbopump. Where they got them, nobody knows. Jerry would keep plugging away at every little lead until he got all the parts. He kept dumping them in my garage until we had enough to start work. I went with Jerry to pick up the bulk of the XLR-25 parts. Some guy had bought it in a junkyard and was planning to put it in a racecar. He and another fellow had invested $700. Jerry and I went to his house in a pickup to get the engine, but there were many key parts missing. Jerry had sources, knew a lot of people, and finally came up with parts enough to make a complete engine. He was a great guy for being persistent and researching."

The X-19 was the last aircraft built by Curtiss. (Courtesy of Emerson Watson)

Em's involvement with the Tornado restoration is extensive. When it was being tested at Caldwell, New Jersey in 1942, Em helped unload it when it came from the factory. He recalls, " There was a fellow from Hughes Aircraft always on hand when the engine ran. The test stand was specially built for the Tornado and was between the parking apron and the hangar. The engine ran with regular flight prop. It ran so smoothly you couldn't tell when a bank was out, but it didn't run a whole lot."

Em went to Connecticut with Jerry to inspect the engine. He worked on the engine with Jerry and Kenny three to four hours per week. He restored many of the accessories and the Curtiss Electric propeller now installed on the engine.

Dave Way

David S. Way was born on March 8, 1922 in Wilmington, Delaware. His family moved to Woodstown, New Jersey when he was three. It was here that he grew up. He graduated from George School and Swathmore College, both near Philadelphia, Pennsylvania. He was drawn to aviation as a result of publicity associated with Charles Lindbergh's flight from New York to Paris, and accounts of the Hindenburg, Macon, and Akron airship crashes.

Dave Way prepared the exploded views and cross section drawings for the Tornado display. (Courtesy of the New Jersey Aviation Hall of Fame and Museum)

After receiving his degree in Mechanical Engineering, Dave's rise at Curtis-Wright is classical. He came to work in 1943 as an engine tester, actually handling the controls and taking data when engines were under test. Next, he worked as a test engineer, making sure that what designers asked for was actually accomplished when the engine was assembled and tested. Special set-ups were sometimes required, and these too were the responsibility of the test engineer. There was a separate instrument group called Test Equipment Engineering. Dave was drawn to this area and became a test equipment engineer. Test equipment expanded while he was part of it because many new test stands were being set up to take care of new engine designs and concepts such as gas turbines and ram jets. During development, it was desirable to test each of the major engine components (compressors, turbines, burners, etc.) independently. Dave then got into the project engineering group where complete engines were developed to the point they would pass qualification testing.

He then worked on actuators for propellers, control surfaces, doors, and landing gears in the Curtiss-Wright propeller plant at Caldwell. When the nuclear vessels program started, Dave went back to Wood-Ridge to participate in the production of pressure vessels, steam generators, and other equipment necessary for nuclear submarine propulsion.

Dave and Jerry worked together on various projects at Curtiss-Wright for many years. When the Tornado arrived in New Jersey, Dave became involved in the project. Dave was not a mechanic by hobby or trade, but he knew how to get things. Dave coordinated activities, chased parts and located materials while the other guys did the actual restoration work. Once the engine was restored, the problem remained of how to display it so its unique features were highlighted. Dave solved this problem by making some posters that showed the internal workings of the engine. He recalls, "We really didn't have the capability to remake the drawings. We had the longitudinal and cross sections from NASM, along with some photographs of parts. I cut and pasted multiple copies of the photographs to produce a sort of exploded view of the engine crankcase."

Augie Zoll

Augie Zoll restored the magnetos for the Tornado. (Courtesy of the New Jersey Aviation Hall of Fame and Museum)

August H. Zoll was born on August 4, 1924 in Brooklyn, New York. He grew up in Glendale, Queens and attended New York City schools. After serving with the Tenth Mountain Division in Italy during WWII, Augie received a degree in Mechanical Engineering from Purdue University in 1948. He joined Wright Aero as an engineer immediately following graduation. It was during this period that Augie met Jerry.

The year 1949 was a big year for Augie. He met his wife, Adie, at Wright Aero. They were married in 1952. Their son, Gregory, was born in 1955. Augie also started taking flying lessons at Teterboro Airport in 1949, but soon after his first solo, ran out of money and had to stop flying.

While at Wright Aero, Augie was responsible for ignition and carburetion for the R-1300, R-1820, and R-3350 engines. He then became involved with Curtiss-Wright ramjets and gas turbine fuel controls. Ultimately he became manager of the Controls and Accessories department. Augie patented the JET-AIR compressor that was used to provide compressed air for snow making at the Great Gorge ski area in Vernon, New Jersey.

During the mid to late 1970s a number of promising energy technologies were aggressively funded by the U.S. Department of Energy. One of these was the use of pressurized fluid bed technology to burn high-sulfur coal in an environmentally acceptable manner. Curtiss-Wright was active in this research which culminated in the construction of a 60-megawatt demonstration plant in 1984. Augie was heavily involved in the design of control systems for this plant. Unfortunately, politics intervened and the Curtis-Wright project was abandoned before it could be started up.

When Wright Aero closed, Augie continued to work for the Rotary Engine Division of John Deere Technologies as an accessory engineer until 1988.

During the move of the exhibit from the Paterson Museum, Jerry began to call on old friends for help. One was Augie. Making use of his experience with engine accessories, Augie restored the seven magnetos for the Tornado.

After Jerry died, Augie, his wife Adie, and a part-time typist, Olga Anotol, continued the process of cataloging the Curtiss-Wright Archive at AHoF. This activity was made possible by donations of computer equipment and partially funded by a grant from *Air & Space* magazine. This activity is still progressing and is resulting in a computerized index of all the records and books in the AHoF Library.

The Real Work of Restoration Begins

During the last days of November 1989, Kenny Bell started building spark plug boards, parts shelves, and labeling parts cans in preparation for doing real work on the Tornado. [12] Edith Abbamont recalls, "His garage was as clean as anyone's kitchen. Everything was set out and organized perfectly." On December 12, 1989, Jerry, Kenny, Emerson Watson, and Dave Way went to NEAM to pick up the engine. It was taken to the garage of Kenny Bell where the remainder of the restoration was accomplished. Open ports were plugged and the magneto booster coil removed from "A" bank. The engine was gone over with compressed air to remove years of accumulated dust and debris. It was then washed down with soap and water and dried with compressed air. Two mouse houses were removed. All hardware slated for removal, including hose clamps, was treated with penetrating oil. The front of the engine was supported and the propeller shaft support was removed and straightened.

On December 16, 1989, the oil drain plugs were removed and the residual oil collected. It appeared to be new Texaco 1200 oil. While the left coolant impeller showed signs of coolant spots, it was noticed the right coolant impeller was new and that the locking tab had not been bent. While the specification sheet called for a Stromberg PR-52, the one on the engine was a PR-55A1. The carburetor intake screen was removed and straightened. Clean preservative oil was found in the fuel system.

On December 29, 1989, the complex process of removing the rear accessory cover began. Having no maintenance manuals, the restoration team had to create disassembly procedures as they went along. Accessories had to be removed in a specific order. There were hidden fasteners behind the carburetor gasket and oil screen housing. A log of all disassembly operations was kept to aid in reassembly. The rear cover was removed on January 2, 1990, and the restorers were delighted to find that the accessory section interior and gears looked like new.

When removal of the ignition harness began on December 19, 1989, many of the spark plug cooling shrouds were impacted with dirt and paint. The material was very hard and had to be carefully removed with a dental pick and vacuum. Additionally, many of the ignition cigarettes were stuck and required treatment with penetrating oil before they could be removed from the spark plugs. This tedious process continued for weeks. As stuck high-tension leads were finally freed, individual magnetos and the primary magneto harnesses could be removed. The spark plugs appeared to have never been used. As they were removed, automotive spark plugs were substituted to prevent debris from entering the cylinders through the spark plug holes. As harness clamps were removed, detailed drawings of the clamp positions and safety wire installation were made. These would be used later to assure an authentic reassembly. Some spark plugs were so corroded that vinegar was used to dissolve the corrosion products, allowing removal of the plugs. The last of the ignition harness components and spark plugs were removed on January 26, 1990.

Once ignition components had been removed from each bank, the valve covers could be removed. Each valve cover was secured by no less than fifty lock wired cap screws. Drawings were made of the lock wire patterns to aid in reassembly. Some condensation was found in the top valve covers. Otherwise, the valves and cams were in good condition. The quantities of oil that had pooled inside the valve spring recesses indicated that residual oil was make-up oil and that the engine had not been run since its assembly.

During the last week of January 1990, the supercharger housing and all intake manifolds were removed. It was not possible to see a great deal of the inside of the engine. Cylinder walls and valves looked new and well oiled, with no corrosion. Delighted with this result, Jerry made a video tape showing what had been done thus far and sent it to Harvey Lippencott. In his February 13, 1990 cover letter, Jerry sent the following preliminary assessment:

> "Based upon my experience with the operation of the Assembly and Test Department of Curtiss-Wright, we have a new engine. The damaged magneto due to a broken cover appears to be new (none of the others have been inspected). The carburetor appears to be used. The engine is so well preserved with a long-term preservative used in that era that at this time no attempt will be made to rotate the engine or will we disassemble any component that was involved in the slosh process."

In April of 1990, Bruce Scott was again called in to consult on the lubricants and preservatives used in the engine. He issued his report on April 4, 1990:

> "It is my opinion that the lubricant/preservative in the engine is the standard preservative oil in use at the time of assembly. This would be Wolf's Head 8-120 plus a preservative additive from E. F. Houghton and Co., Philadelphia, PA that met the requirements of AN-VV-C-576."
>
> "If any re-preservation is to be used, I suggest using the material drained from the engine (which may be depleted in preservation capability due to 'plating out' of the inhibitor on the metal surfaces) or contacting Houghton for a small supply of the AN-VV-C-576 inhibitor (a gallon should do) and add that in proper proportion (Houghton can tell you) to a straight parafinnic base mineral oil (Wolf's Head 8-120, Texaco 1100 or 1200, or equivalent)."
>
> "While the SP-400 is compatible with engine lubricants (a requirement of MIL-C-16173 so that it can be used with pre-packaged roller and ball bearings, etc.) it would be difficult to coat the engine interior by the necessary flow methods. In addition the solvent vapors of 1,1,1 Trichloroethane and mineral spirits would be trapped inside which could have a deteriorating effect on non-metallic components and even on metal surfaces. I would have no hesitancy to use the SP-400 on any accessible surfaces when it can be sprayed or applied by brush..."
>
> "As I have mentioned in our previous discussion, the material coating the inside of the ignition "cigarettes" is typical of this [paralketone] product."

Once the engine was stripped of magnetos, ignition harnesses, accessories, and manifolds, actual restoration could begin. The ignition harness cooling shrouds were first. Years of abuse and neglect had taken their toll, and each of the fourteen shrouds had to be straightened. They were then cleaned and painted with dull aluminum paint followed by a clear coat.

After having been cleaned and painted, ignition shrouds hang from the ceiling of Kenny Bell's garage. Plastic bags protect the distributor blocks. Intake shrouds are in the foreground, exhaust shrouds, with their closer spark plug spacing, are in the background. (Courtesy of David S. Way)

All open mounting pads were covered with duct tape as were fittings, intake ports, and exhaust ports. All of the valve covers were temporarily reinstalled to keep debris out of the engine. The propeller shaft had undergone years of neglect, so several applications of Naval Jelly, plus considerable manual effort was necessary to remove the rust.

Now that the engine had been disassembled and prepared for a deep cleaning, the restoration team took stock of the engine and reached a definite conclusion that the engine had never run. The following reasons were cited:

- Coolant ports in the supercharger housing were clean
- Coolant pump impellers looked new and exhibited no signs of coolant
- Coolant inlet and outlet ports were clean
- Nose case gears showed no sign of use
- There was no carbon in the exhaust ports
- Neither inlet nor exhaust valves showed any sign of use
- Intake manifolds showed no sign of air flow
- Spark plugs showed no sign of use
- Supercharger impeller exhibited no air flow patterns
- Inside of the cylinders were clean, oiled, and showed hone marks
- Rear accessory gears showed no sign of use
- No oil was in the oil pump inlet fitting
- Engine was loaded with build-up and preservation oil

A boroscope was borrowed and used to examine each of the 42 cylinders. All were in great condition. A check was also made of the crankcase interior. It too was in original condition.

The task of restoring the individual parts was a monumental one. Each part, whether it was as large as the accessory case or as small as a nut or washer, had to be cleaned and painted. Repair was sometimes required. The restoration activity log has months of details about how the individual parts were cleaned and painted. Nearly all of this work had to be done without the aid of power tools. There was a lot of scraping, use of steel wool, washing in solvent, and blow-drying. Kenny's February 17, 1990 log entry is a good example of this: "Scraped, sanded, steel wool, primed and painted another set of [ignition harness cooling] shrouds. It takes approximately 5-6 hours to do one set from start to finish (there were fourteen sets to do). Wash, straighten, scrape, sand, steel wool, prime, paint, and clear coat. This does not include the time to repair the vacuum cleaner and clean up!" Then, on February 22, "Inspected the exhaust shrouds I have done. I will have to do one set over – it does not meet my standards. I must have been dreaming!" Such it was for many of the parts. Difficult, mind-numbing work that might have to be repeated if the final result fell short of the high standards the restoration team had set for themselves. Edith Abbamont remembers that Jerry did all of the valve covers in their back yard. "He would scrape and fill and paint. If it wasn't perfect, he would scrape it all off and start over."

Of particular difficulty was the engine truck. The engine truck had been designed to allow rotation of the engine, providing easy access to all engine components. To achieve this, the engine had been attached to a large ring via four mounting lugs. Multiple rollers supported the ring within the engine truck frame. Unfortunately, the mounting lugs were too long and contacted the rollers as the engine was rotated. Additionally, the engine had apparently been dropped at some point and the upper left mount had bolts sheared. To correct this, restorers had to jack the engine, remove each lug, extract any sheared bolts, and rework it to clear the ring rollers. Before it was reinstalled, each lug was cleaned, filled, primed, and painted. Removal of the lugs was no easy task. Fasteners that had been exposed to the elements for better than 40 years required large quantities of penetrating oil. Later, when the mount was prepared for painting, all traces of penetrating oil had to be removed. The ring rollers got similar treatment. Since it was impractical to remove the engine from the truck, all cleaning, repair, priming and refinishing had to be done with the engine in place or suspended on jacks.

Since no parts catalog was available for the Tornado, Jerry and Kenny used a Wright Cyclone 18 parts catalog to obtain the part numbers for gaskets and fasteners common to the accessories of both engines. New gaskets and fasteners were acquired from sources of Wright Aeronautical surplus stock. Many gaskets were unique to the Tornado. Replacements for these were painstakingly cut from gasket material.

All through the restoration process, members of the team would take certain parts from Kenny Bell's garage and work on them at their own house. There is no record of a single part ever being lost, however a few were "temporarily missing."

Most of the stripper, primer, and paint was donated by Randolph Products of Carlstadt, New Jersey. Randolph had supplied the paint used when the engine was originally finished. The following paints were used:

For the ignition harness – Dull Aluminum followed by clear acrylic;

For the baffles, cylinder banks, other black parts – Randolph No. 303 Engine Black;

For the Exhaust port dust covers – Scarlet Red Krylon No. 2116;

For the Crankcase – Battleship Gray.

Work on the cylinder banks and nose case had to be done on the engine, and therefore, by hand. There was much work with a wire brush and scraper done by Jerry and Kenny during the Fall of 1990. When the laborious cleaning process was complete, the blocks were painted black and the case gray. Like the cleaning, this painting had to be done by hand with a small artist's brush. Because of the difficulty in rotating the engine, all cleaning, preparation, and painting was accomplished on one block before the engine was rotated to allow better access to another.

As specified by the development contract, the Tornado had been delivered to Wright Field without any accessories. As the restoration proceeded, it became evident that accessories would be a nice addition. On June 6, 1990, Jerry wrote a letter to Associated Products of America requesting the following:

Part	Vendor	Model	Type
Starter	Eclipse	M-15	602
Generator	Eclipse	AC-021	16 tooth
Fuel Pump	Pesco	CWB	R-60
Vacuum Pump	Pesco	B-8	211-J
Hydraulic Pump	Vickers	3.5 x 3.5, 4 hole	
Tachometer Gen.	GE. or Kollsman		

These items were furnished by Associated Products free of charge, as was a complete Curtiss Electric 3-blade propeller. This was the small-shank C-63-S design.

Em Watson cut down the propeller blades to a size that could be accommodated by the exhibit and restored the pitch change mechanism to operating condition. Em explains, "Since the slip rings were in bad shape, they were removed and the prop is hard-wired. You can change pitch from stop-to-stop, mainly so kids can see it operate. Jerry took care of getting the box that controls this and provides 24-volt power to it." Em also restored the engine accessories. Since Jerry wanted the completed engine to be as light as possible, all the internal parts were removed from the starter and generators.

When the job of refurbishing the Tornado magnetos was identified, Augie Zoll was the natural choice. His work with accessories during the early part of his career at Curtiss-Wright gave him the knowledge and connections to get the job done. He removed all the magnetos from the engine, disassembled them, cleaned and inspected the parts, painted the exterior, reassembled the magnetos and reinstalled them on the engine. Aside from a lot of dirt and some minor surface corrosion to the housings, the biggest job Augie faced was finding duplicate data plates. The original ones were so badly damaged by corrosion that they could not be reused. Augie wrote the Bendix Engine Products Division, requesting its assistance in locating duplicate data plates stamped with the original information. Bendix was able to locate enough data plates for the seven magnetos and fourteen ignition harnesses in Australia. The data plates were obtained and stamped with original numbers from Bendix records.

Finally the long process of cleaning began to wind down. One particularly vexing problem remained. That was removing caked heavy grease and grime from the ignition harnesses and distributor caps. It was finally found that multiple application of oven cleaner followed by trichlorethane did the job. In October of 1990, Emerson Watson often joined Jerry

One example of lock wire on the rear accessory housing plugs.

and Ken in cleaning activities on the engine itself. While there was still cleaning and painting to be done, emphasis shifted to putting the project back together. Gaskets were made, and plans were made for the long assembly process that lay ahead.

In late November of 1990, the team began to actually put parts back on the engine. As each assembly was completed, safety wire was installed according to the drawings made during disassembly. Mel DeBlock did much of this long and tedious process. His efforts resulted in a beautiful presentation that is one of the high points of the engine today.

Before the engine transfer was ever accomplished, NEAM agreed to foot the bill for the plating of hardware.[13] Jerry went to pick it up on November 13, 1990. Most of the visible hardware had been cadmium plated. In some cases, a tap had to be run through the threads to remove excess plating.

On December 11, 1990, almost one year after the initial receipt of the engine, a group of Wright Aeronautical Corporation retirees gathered to inspect the progress. Present were Bill Johnson, Jonas Anderson, Fred Chamberlin, Earl Mathes, Emerson Watson, Angelo Defeo, Augie Zoll, Rossa Cole, Dave Way, Bruce Scott, Jerry Abbamont, Kenny Bell, and Bill Litke. Once the restoration inspection was complete, the group adjourned to the pub for lunch and reminiscing that lasted the remainder of the afternoon. It was reported that a good time was had by all.

Jerry and the team planned to install a Curtiss Electric propeller on the Tornado when it was displayed, rendering the original propeller shaft brace unusable. Instead, a brace from the engine to the stand was constructed of pipe. This was placed underneath and essentially out of sight. A cover for the engine stand was constructed from plywood. Planning began for the process of moving the engine from the garage to the museum.

The process of engine assembly continued until March 11, 1991 when the last of the accessories were installed. The engine was then moved to the Aviation Hall of Fame and Museum where Em Watson and Kenny Bell installed the newly refurbished propeller on the engine. Jerry and Dave Way had been concentrating on display of the Tornado. Dave had done many of the placards that explain the significant features of the engine, while Jerry had busied himself with an audio presentation. It is Jerry's voice on the audio that explains the Tornado exhibit. With these finishing touches by Jerry, Dave, Em, and Kenny, the Tornado exhibit was complete. After decades of searching, years of negotiating, and still more years of restoration, the prodigal Tornado had finally returned home to a place of honor in New Jersey.

The agreement between NEAM and AHoF that got the Tornado to New Jersey included a reciprocal loan by AHoF of a Curtiss-Wright RC-260-Y8 rotary combustion (Wankel) engine. Curtiss-Wright, in one of its few internally funded developments after World War II, had poured enormous amounts of money into the development of this power plant concept for aircraft use. The world-renowned engine consultant Max Bentele had been a key player in the project. In the end, the rotary combustion engine did

The Curtiss Electric propeller was restored by Emerson Watson. Here, Em and Kenny Bell are installing the propeller. (Courtesy of the New Jersey Aviation Hall of Fame and Museum)

View of the Propeller with the Pitch Change Motor Installed (Courtesy of the New Jersey Aviation Hall of Fame and Museum)

Opposite Page and Above: The Tornado Today

not provide enough of a competitive advantage to warrant production, and the rights were sold to John Deere Technologies.

The RC-260-Y8 was one of the experimental engines resulting from the Curtiss-Wright development effort. This engine was sent to NEAM on January 28, 1991.[14] In December of 1997, NEAM and AHoF agreed to make the engine trade permanent.[15]

A Tribute to the Restorers

The Tornado now sits in a place of honor in the Aviation Hall of Fame and Museum of New Jersey. In addition to being an important piece of technological history, it is a tribute to all associated with its restoration.

Jerry Abbamont had the drive and determination to make it happen. He organized, led, and inspired a team of skilled craftsmen, each of whom contributed enormous amounts of time to the project. For Jerry and Kenny Bell, whose time was running out, the Tornado was their last project. For all of the other restorers, the Tornado is a fitting reminder of their passion for engine history, their fondness for the people at Curtiss-Wright, and their dream that some of it would be remembered.

---

1 This was related to Graham White by Harvey Lippencott, electronic mail from Graham White to the author, December 3, 1999.

2 Herbert Driggers, letter to Robert B. Meyer, Jr., NASM Curator of Propulsion, November 6, 1974. See also correspondence between George Surasky and Robert B. Meyer, Jr, December 27, 1977, January 2, 1978, August 4, 1978, January 23, 1979 and January 25, 1979. See also correspondence between Robert B. Meyer, Jr. and Col. Richard Uppstrom, Director, U.S. Air Force Museum, April 20, 1978 and August 17, 1978, *Correspondence Regarding R-2160 Restoration.*

3 Telephone conversation between Mrs. Walter Soplata and the author, March 19, 2000.

4 Harvey Lippencott, letter to Jerry Abbamont, March 23, 1990, *Correspondence Regarding R-2160 Restoration.*

5 Harvey H. Lippencott Obituary, *The New York Times* (January 3, 1997).

6 Unless otherwise noted, all material in this section comes from telephone interviews with the restoration team conducted by the author during February through December, 2000.

[7] Unless otherwise noted, all material in this section comes from *Correspondence Regarding R-2160 Restoration* and *Restoration Log*, as well as telephone interviews with the restoration team conducted by the author during February through December, 2000.

[8] H. V. Pat Reilly, letter to Robert North, Curator, New England Air Museum, September 3, 1987, *Correspondence Regarding R-2160 Restoration*

[9] R. L. North, letter to Pat Reilly, September 16, 1987, *Correspondence Regarding R-2160 Restoration.*

[10] David G. MacMillan, Trustee, Letter to Mike Speciale, Museum Executive Director, New England Air Museum, April 21, 1988, *Correspondence Regarding R-2160 Restoration.*

[11] Unless otherwise noted, all material in this section comes from telephone interviews with the restoration team conducted by the author during February through December, 2000.

[12] Unless otherwise noted, all material in this section comes from *Correspondence Regarding R-2160 Restoration* and *Restoration Log*, as well as telephone interviews with the restoration team conducted by the author during February through December, 2000.

[13] Jerry Abbamont, letter to Harvey Lippencott of the New England Air Museum, October 17, 1989, *Correspondence Regarding R-2160 Restoration.*

[14] *Restoration Log*, 72.

[15] Mike Speciale, NEAM Executive Director, letter to the author, November 28, 2000.

8 Conclusions

The sole surviving Tornado is a credit to those who designed it, built it, and restored it.

What if?

It is tempting to predict the future with the benefit of 60 years of hindsight. It is hard to examine any human endeavor, especially one as complex as the Tornado without wondering "what if?" What if the timing of World War II had been different? What if Wright Aeronautical had possessed the talent, time, and motivation to complete development of the Tornado? Would it have been successful?

There can be little doubt the engine could have been made to work. By all accounts, most of the major problems had been solved when development ceased. There may have been other difficulties lurking that would have shown up as engines accumulated more time. For example, there are the unconfirmed comments of George Suraski: "...based on the experience I had with the 42-cylinder engine I doubt it would have been a viable engine. The layshafts connecting the 14-cylinder sections to the reduction gear in the front of the engine were rather slim. As the engine was mounted, there was measurable sag from front to rear due to engine weight. This caused an undesirable whip in the layshafts, presenting future potential failures."[1]

If this were true, then it was a design problem that may have proved insurmountable, particularly with the planned 56 and 70-cylinder versions of the engine. However, in the thousands of pages reviewed while researching this book, only Suraski mentions any problems with engine sag or layshaft whip. This, however, does not mean that problems with engine length did not exist. Wright Aero intended to mount the engine with seven members attached to the rear face of each cylinder bank. This put enormous bending loads on the engine and mount. Pratt & Whitney chose a similar mounting scheme for its early R-4360. It had to be changed twice, each time moving the mount further forward, until a satisfactory solution was achieved by mounting the engine near its center of gravity. While the Tornado was stiffer because of its one-piece cylinder banks, engine flexure would have likely been a problem.

Whether the engine would have ever produced the promised 2,500 hp is another issue. Due to the small cylinders and long induction pathways, BMEP was down significantly from the Cyclone 9 upon which Tornado performance estimates were based. If it had ever produced 2,500 hp, it would have probably done so after a redesign of the cylinder head configuration and induction plumbing. According to John Minnich, "after the program was over, Carl Stuart, a Wright Aero test engineer, ran an experimental engine consisting of a single Tornado cylinder bank and crankcase. By using tuned lengths of pipe in the intake manifold, Stuart solved the air distribution problem."[2] W. G. Lundquist's patent number 2,382,244, issued on August 14, 1945, detailed an intake manifold arrangement that also would have eliminated the air distribution problems with the Tornado.

The Tornado would have benefited from a technology that came into common usage during the middle part of World War II. This was anti-detonation injection (ADI), or as it was more commonly known, "water injection." ADI allowed engine power to be increased by as much as 20 percent, and was simple to apply. All that was required was a small water tank, a pump, an injection regulator, and some plumbing. The water tank was filled with water mixed with alcohol to prevent freezing at high altitudes. When additional power was required for take-off or high-speed dashes, the water was injected into the supercharger eye and served to cool the intake charge. By cooling the intake charge, considerably higher manifold pressure could be tolerated without detonation, and engine power was significantly increased.

The Tornado would have almost certainly grown in diameter and complexity because of these changes. It was, by the most generous assessment, a nightmare to manufacture and assemble. It required a number of large and complex precision castings and forgings. Many machining operations, such as line boring of the layshaft bearings and camshaft bearings had to be done on partially complete machining assemblies. This meant that some machining operations were done on parts that were then assembled, line-bored, disassembled, cleaned, painted, and then kept together forever. Assembly of the engine, with the myriad gears and fasteners, was also no trivial matter.

A fully developed Tornado would have also required considerably more testing. When work ceased on the Tornado, the six 42-cylinder engines had undergone a total of 1,631.35 hours of testing. An additional 1,200 hours had been run on two Tornado 14s and about 130 on the six-cylinder test engine. By comparison, the Pratt & Whitney R-2800 had undergone 3,500 hours of testing at the time it completed its type test. An even more telling statistic has to do with the Pratt & Whitney R-4360, an engine approaching the complexity of the Tornado. The R-4360 had 15,000 hours of ground testing on 23 engines before it became reliable. Clearly, the Tornado had a lot of bench running left to do.

The high cost of development for so few engines must have played a part in the outcome of the project. When Wright Aero folded the Tornado program, nearly 6.5 million dollars had been spent. In year 2000 dollars, this would have been about **62.5 million**, a whopping huge amount by any standard. By comparison, the R-2800 cost 8 million and the R-4360 cost 25 million. With the limited market for the Tornado, the amortized cost per engine would have been very high. Even in times of war, there was some limit to the depth of Army pockets. If Wright Aero was to have successfully plundered these pockets, it had to be done with engines in volume production. If the Tornado had gotten into production, there was still no guarantee of a successful application. Wright Aero, with an eye on profits, simply decided to stop throwing good money after bad and to concentrate on engines built by the thousands.

Then there is the issue is how the Tornado would have performed in service. If the service history of the Pratt & Whitney R-4360 and Napier Sabre are any indication, we must assume the Tornado would have been temperamental in operation and would have required a large number of maintenance hours per every hour of flight. The presence of 84 nearly inaccessible spark plugs surrounded by sheet metal cooling shrouds is a terrible image to any technician. The same is true of the 84 valves hidden behind seven covers, each with over 50 lock-wired fasteners. The high-speed layshafts, spinning at over 12,000 rpm, would have made excellent oil separators and accumulated prodigious amounts of sludge. If the engine life exceeded around one hundred hours, this would have to be removed. Otherwise, it would have occluded the small oil passages associated with the layshafts. Then there were the induction hose clamps, the coolant hose clamps, the myriad external oil and coolant lines, the seven magnetos, each of which required individual timing adjustment. None of this paints a pretty picture of the maintenance. But such is the cost of going fast.

Why did the Tornado Program Fail?

There are many reasons that the Tornado never reached its full potential. Some were technical, but most were political and economic. It is true that the new priorities of World War II finished the Tornado, but it was in deep trouble long before the war started.

If blame is to be apportioned, it must rest squarely with Wright Aero. The Army went to great lengths to support the Tornado. The Army forced recalcitrant airframe makers to use the engine. The Army tolerated one schedule slip after another. The Army even bought engines that were unneeded just to make the deal attractive to Wright Aero. All Wright Aero had to do was the job it was contracted for. Yet, while the Tornado languished, Wright Aero was making secret deals with the Navy for development of the brand new air-cooled R-4090 engine.

Wright Aero chose to forge ahead into multi-cylinder development without the requisite single-cylinder work. This must have meant that Wright Aero was confident about the basic cylinder design. With engine luminaries such as Dr. Arthur Nutt of Curtiss D-12 fame leading the engineering team, it seems that no one seriously questioned this decision. However, the perfection of cylinders can be a most difficult part of engine development, and no one, no matter how experienced, is immune to mistakes in this most fundamental step of engine creation. Despite their best intentions, the Wright Aero designers' choice to skip the initial single cylinder phase would cost them dearly as the most basic types of cylinder construction difficulties showed up in the multi-cylinder Tornados.

Two kinds of cylinder problems—valve component breakage and manifold insufficiency—introduced delays into the program. Valve component breakage should have been anticipated. Nearly every high-power aircraft engine ever built has suffered through a phase of valve development. Tornado cam profiles required tweaking and weak components required redesign. Single cylinder testing would have identified valve problems early and ended them. It is vastly simpler to rebuild a single cylinder test engine that has trashed itself after a valve failure than a 42-cylinder engine that is built like a Swiss watch. The second kind of problem, that of manifold insufficiency, could have not been anticipated as a result of single cylinder tests. However, by eliminating valve problems before hand, diagnosis and treatment of the manifold issues would have been easier. It is also noteworthy that Wright Aero's reluctance to build a dedicated experimental machine shop was also a delaying factor. Dependence on outside vendors for experimental parts injected huge delays into the program.

Paradoxically, it should be noted that if Wright Aero had proceeded in a more traditional way, the extended development cycle might have been so unattractive that the program could not have been sold to the Army in the first place.

With all this blame for the Tornado's failure being placed on Curtiss-Wright management, it is interesting to explore the corporate political climate and corporate culture at Curtiss-Wright. The managers blamed the technical people and the technical people blamed the technology. Guy W. Vaughan, in a March 26, 1949 letter to Paul V. Shields, who succeeded him as President of Curtiss-Wright, was very critical of Dr. Arthur Nutt. Vaughan wrote:

> "At the time Curtiss and Wright were merged, Arthur Nutt was forced on Wright Aeronautical

as Vice President of Engineering by Dick Hoyt (Chairman) because he had an international reputation based on the performance of the D-12 engine, which, incidentally, was designed by an engineer named Kirkham, and not by Nutt as supposed."

Regarding the Tornado:

> "A 42-cylinder liquid cooled engine was designed and a prototype model authorized. This engine was known as the "Tornado" or "Corn-cob" and had very interesting specifications. Valve trouble was responsible for the engine's ultimate failure. Nutt was urged to get a valve specialist by the name of Miller who designed engines for the Indianapolis speedway that ran as high as 7,200 rpm without valve difficulties. However, Nutt refused to call him in until it was too late and the engine never qualified for production.[3]

Arthur Nutt had a predictably different view of the Tornado:

> "The engine with its six crankshafts geared into the propeller reduction gear was far too complicated and full of torsional vibration problems. It never completed its experimental stage and never went into production."

This is the only mention of any problems with torsional vibration and drive train components. One can only assume that Nutt, isolated from the program by his many duties as vice president of engineering, lost track of the technical issues. Also interesting is Nutt's perception of the Curtiss-Wright management team of which he was presumably a part:

> "As mentioned previously I was given this appointment [to Vice President of Engineering] as a result of the Curtiss-Wright merger and naturally this action was not well received by many of the Wright employees. These 14 years, while productive from the standpoint of my association with the development of many new engines and the improvement of Wright engines, were perhaps the most difficult years in my career. I had been thrown with a hostile organization, which I had to persuade to work together. I was given a Chief Engineer who was very capable but nevertheless resented my presence."

> "I cannot close this chapter of my career without commenting that I was quite unhappy in the hostile atmosphere at Wright Aero. I felt that I was not wanted and that many assignments such as foreign travel were given to me to keep me away from the engineering, which I was so fond."

In short, the situation was this: A corporate president distrusted a technical vice president who he did not respect. The vice president, feeling small, overworked and under appreciated, was not motivated. The technical people, without strong leadership and a sense of urgency from upper management, did the best they could with what they had. They were probably happy to move to the next new thing when the real work of sorting through the last one percent of the problems began.

Curtiss-Wright, with its investment-banking heritage, was not an environment where new concepts like the Tornado could flourish. While this philosophy was certainly bad for the Tornado, bad for the country, and ultimately bad for Wright Aero, no one can argue it was bad for profits. By avoiding risk and concentrating on profitable production, Curtiss-Wright finished the war as the second richest corporation on earth. Yet, in the end, this very philosophy fueled the downfall of the company. When the profit taking was over, so was the engineering staff and development. Only a second-tier job shop remained.

Which is best – sleeve valves or poppet valves?

The Tornado was meant to be the best engine possible using poppet valve technology. It produced a respectable 2,350 hp from 2,160 cubic inches but weighed a disappointing 2,700 pounds. The real question is how the Tornado compared to rival engines with sleeve valves.

The Army got into the Hyper engine project in the first place because of a debate over sleeve verses

This beautiful Napier Sabre cut-away from the collection at RAF Hendon produced 3,055 hp from 2,400 pounds.

poppet valves. On the other side of the pond, several British concerns, notably Bristol, Napier, and Rolls-Royce had soldiered on in the cause of sleeve valve engines. Perfection of this technology came at an appalling cost of manpower and money, but it was done. Thousands of material, process, and tooling combinations were tried before acceptable sleeves could be mass-produced. By the end of World War II, the Napier Sabre VII, an H-24 engine, was producing 3,055 hp from a displacement of 2,239 cubic inches and a weight of around 2,400 pounds. This engine was powerful enough to allow the Hawker Tempest to chase down and destroy Nazi V-1 "Buzz Bombs." [4]

The crown jewel in sleeve valve development was the Rolls-Royce "Crecy." It was a revolutionary highly supercharged two-stroke liquid-cooled V-12 that produced 2,750 hp from 1,536 cubic inches and a weight of 1,650 pounds! Versions on the drawing board were expected to ultimately produce over 3,500 hp. In addition to impressive specific power and specific weight, the Crecy had very good specific fuel consumption. While it suffered its share of development troubles, most had been solved, and solutions were in the works for the remaining ones. Like the Tornado, the Crecy was abandoned when World War II caused higher priorities to be placed on other projects. [5]

The Sabre and Crecy have far better specifics than U.S. poppet valve engines. This is mainly because sleeve valve engines can tolerate considerably higher boost pressure than poppet valve engines. While the exhaust port of a sleeve valve remains relatively cool, parts of a poppet exhaust valve run at a red heat and contribute to detonation at high boost pressures. One is forced to wonder what the U.S. could have done with the sleeve valve had Sam Heron not intervened.

The Rolls-Royce Crecy was designed to produce 3,500 hp from 1,650 pounds. (Courtesy of Rolls-Royce Heritage Trust)

Requiem

Like many legends, the Tornado died young, before its full potential could be realized. Like many legends, we remember it more as an object of art than as a practical machine. It never flew, never passed a type test, and never went into production. However, as a thing of beauty, it does great credit to those who designed it, those who built it and those who restored it. The Tornado was for Wright Aero the final chapter in a long history of going fast. It was Wright Aero's last liquid-cooled reciprocating engine.

---

1 George Surasky, letter to Robert D. Meyer, Jr., Curator of Propulsion, NASM, January 25, 1979, *Correspondence Regarding R-2160 Restoration.*.

2 John Minnich, telephone interview by the Author, December 2000.

3 Guy W. Vaughan, Letter to Paul V. Shields, March 26, 1949, (AHoF Archive).

4 White, *Allied Aircraft Piston Engines*, 162-179.

5 Nahum. *The Rolls-Royce Crecy.*

Appendix A—Design Studies of Liquid-Cooled Engine Types

(From *Summary Report of Flat and Radial Liquid Cooled Engine Design Studies*, Wright Aeronautical Corporation Serial No. 436, August 30, 1939)

Table A.1. Desirable Features of Four Candidate Engine Configurations

TYPE	NUM CYLS	DISP (cu in)	DESIRABLE FEATURES
Opposed	12	1920	Simplicity Maximum horsepower per inch (120 hp/in, 15.0 inches high) Small number of cylinders Allows combining two units for 3600 hp
H	24	1820	Short crankcase length Rigid crankcase Small bore (4,625") Low BMEP (183 for 1800 hp) Smooth operation Output can be increased to 2400 hp with 32 cylinders Two engine units can be combined for higher power Half engine unit will make compact 12 or 16 cylinder opposed engine with 900 to 1200 hp High horsepower per inch of height (86 hp/in, 21.0 inches high)
Flat X	24	1820	Short crankcase length Rigid crankcase Small bore (4.625") Low BMEP (183 for 1800 hp) Smooth operation High power per inch of height (100 hp/in, 18.0 inches high Output can be increased to 2400 hp with 32 cylinders Two engine units may be combined for higher power Half Engine unit will make compact 12 or 16 cylinder flat vee with 900 to 1200 hp
Barrel	24	2020	20.5 inch diameter Simplicity of design Low BMEP (174 for 1800 hp) Favorable induction system design Half engine unit will make 900 hp 12-cylinder engine

Table A.2 Undesirable Features of Four Candidate Engine Configurations

TYPE	NUM CYLS	DISP (cu in)	UNDESIRABLE FEATURES
Opposed	12	1920	Length Lack of vertical crankcase stiffness High BMEP (207 for 1800 hp) Split connecting rod bearings Large cylinder bore Poppet valve gear design for 3570 rpm
H	24	1820	Two crankshafts Difficulty in gearing two crankshafts to propeller shaft Multiplicity of parts Weight Requires 90 degree propeller shaft drive for DB10A Two valves per cylinder Split connecting rod bearings Poppet valve gear design for 4260 rpm Possible inter-cylinder mixture distribution problems Complicated ignition compared to 12-cylinder engine
Flat X	24	1820	Unconventionall connecting rod system Split connecting rod bearings Two valves per cylinder Multiplicity of parts Poppet valve gear design for 4260 rpm Possible inter-cylinder mixture distribution problems Output can be increased to 2400 hp with 32 cylinders Complicated ignition compared to 12-cylinder engine
Barrel	24	2020	Crankcase length (approximately 12 feet overall length) Experimental enginetype

Table A.3 Comparisons of Opposed Liquid-Cooled Engines

Number of Cylnders	**8**	**12**	**16**	**12**	**16**	**12**
BHP	900	900	1200	1060	1440	1800
RPM for piston speed of 3200 fpm	3570	4260	4260	4150	4150	3570
Bore (inches)	6.12	4.62	4.62	5.00	5.00	6.12
Stroke (inches)	5.37	4.50	4.50	4.62	4.62	5.37
Displacement (cu in)	1275	910	1215	1093	1460	1920
BHP/cu in/min.	0.000263	0.000234	0.000234	0.000234	0.000234	0.000263
Brake Mean Effective Pressure	207	183	183	188	188	207
Height (inches)	16	15	15	17.0	17.0	16
Width (inches)	48	44	44	45.0	45.0	48
Crankcase Length (inches)	34	36	36	38.2	52.0	48
Valves per cylinder	4	2	2	2		4
Valves per engine	32	24	32	2	2	48
Displacement per cylinder (cu in)	159	75.8	75.8	90.8	90.8	159
Piston Area (sq in)	237	202	259	236	315	35

Table A.4 Comparisons of H-Type Liquid-Cooled Engines

Number of Cylnders	**24**	**24**
BHP	3600	1800
RPM for piston speed of 3200 fpm	3570	4260
Bore (inches)	6.12	4.62
Stroke (inches)	5.37	4.50
Displacement (cu in)	3840	1820
BHP/cu in/min	0.000263	0.000234
Brake Mean Effective Pressure	207	183
Height (inches)	32	21
Width (inches)	48	44
Crankcase Length (inches)	48	36
Valves per cylinder	4	2
Valves per engine	96	48
Displacement per cylinder (cu in)	159	75.8
Piston Area (sq in)	710	404

Table A.5 Comparisons of X-Type Liquid-Cooled Engines

Number of Cylnders	**24**	**24**	**32**	**32**
BHP	1800	2160	2400	2880
RPM for piston speed of 3200 fpm	4260	4150	4260	4150
Bore (inches)	4.62	5.00	4.62	5.00
Stroke (inches)	4.50	4.62	4.50	4.62
Displacement (cu in)	1820	2185	2425	3000
BHP/cu in/min.	0.000234	0.000234	0.000234	0.000234
Brake Mean Effective Pressure	183	188	183	188
Height (inches)	18	21.0	18	21.0
Width (inches)	44	45	44	45.0
Crankcase Length (inches)	36	38.2	36	52.0
Valves per cylinder	2	2	2	2
Valves per engine	48	48	64	64
Displacement per cylinder (cu in)	75.8	90.8	75.8	90.8
Piston Area (sq in)	404	472	538	620

Table A.6 Comparisons of Double-X-Type Liquid-Cooled Engines

Number of Cylnders	**48**	**48**	**64**	**64**
BHP	3600	4320	4800	5760
RPM for piston speed of 3200 fpm	4260	4150	4260	4150
Bore (inches)	4.62	5.00	4.62	5.00
Stroke (inches)	4.50	4.62	4.50	4.62
Displacement (cu in)	3640	4370	4850	6000
BHP/cu in/min.	0.000234	0.000234	0.000234	0.000234
Brake Mean Effective Pressure	183	188	183	188
Height (inches)	36	42.0	36	42.0
Width (inches)	44	45	44	45.0
Crankcase Length (inches)	36	38.2	48	52.0
Valves per cylinder	2	2	2	2
Valves per engine	96	96	128	128
Displacement per cylinder (cu in)	75.8	90.8	75.8	90.8
Piston Area (sq in)	808	945	1077	1260

Table A.7 Comparisons of V and Barrel-type Liquid-Cooled Engines

Configuration	**V**	**V**	**V**	**V**	**Barrel**
Number of cylinders	12	12	16	16	24
BHP	900	1080	1200	1440	1800
RPM for piston speed of 3200 fpm	4260	4150	4260	4150	4040
Bore (inches)	4.62	5.00	4.62	5.00	4.75
Stroke (inches)	4.50	4.62	4.50	4.62	4.75
Displacement (cu in)	910	1093	1215	1460	2020
BHP/cu in/min.	0.000234	0.000234	0.000234	0.000234	0.000223
Brake Mean Effective Pressure	183	188	183	188	174
Height (inches)	30	31.0	30	31.0	20.5
Width (inches)	18	21.0	18	21.0	20.5
Crankcase Length (inches)	36	38.2	48	52.0	80
Valves per cylinder	2	2	2	2	3
Valves per engine	24	24	32	32	72
Displacement per cylinder (cu in)	75.8	90.8	75.8	90.8	84.2
Piston Area (sq in)	202	236	259	315	456

Table A.8. Power range possible with small diameter, liquid-cooled radial engine design (4.125" bore x 3.500" stroke)

Number of cylinders	**7**	**14**	**28**	**42**	**56**	**70**
Displacement (cu in)	328	656	1312	1968	2625	3280
Initial horsepower at 169 BMEP and piston speed of 2500 fpm	300	600	1200	1800	2400	3000
Developed horsepower at 210 BMEP piston speed of 2870 fpm (1939 G200 performance)	430	860	1720	2580	3440	4300
Approximate weight (lbs)	402	765	1450	2070	2620	3275
Diameter (inches)	33.0	33.0	33.0	33.0	34.0	35.0

Table A.9 Power range possible with small diameter liquid-cooled radial engine design (4.250" bore x 3.625"; stroke, 51.42 cu in cylinder)

Number of cylinders	**7**	**14**	**28**	**42**	**56**	**70**
Displacement (cu in)	360	720	1440	2160	2880	3600
Initial horsepower at 210 BMEP and piston speed of 2500 fpm	394	789	1577	2365	3160	3945
Approximate weight (lbs)	465	888	1680	2400	3040	3800
Diameter	36.5	36.5	36.5	36.5	37.5	38.5
Piston area (sq in)	99.4	198.8	397.5	596/0	795/0	994.0

Table A.10 Design Studies of Various Radial Engine Configurations

Cylinders per Row	Cylinders per Engine	Bore x Stroke (inches)	Stroke/ Bore	Piston Area (sq. ft.)	Angularity with Master Rod (degrees)	Angularity with Kinetic Rod (degrees)	Outside Diameter (inches)	Power @ 210 BMEP and piston speed of 2870 fpm (1939 G200 cyclone 4.53 hp/sq in)
5	30	4.750 x 4.250	.900	532	21.5	Out	37.5	2410
5	30	4.750 x 3.625	.794	532	21.5	17.0	35.0	2410
5	30	4.750 x 3.375	.710	532	21.5	17.0	34.0	2410
5	30	4.750 x 3.625	.764	532	23.0	Out	35..5	2410
6	36	4.375 x 3.500	.800	540	22.0	17.5	33.9	2450
6	36	4.375 x 3.940	.900	540	21.5	17	35.5	2450
6	36	4.375 x 3.500	.800	540	23.0	Out	33.1	2450
6	36	4.375 x 3.375	.770	540	21.5	17.0	33.6	2450
6	36	4.750 x 3.375	.710	637	21.5	17.0	34.5	2850
7	42	4.000 x 3.625	.900	530	21.5	17.0	33.5	2400
7	42	4.125 x 3.500	.848	565	21.5	17.0	33.2	2560
7	42	4.500 x 3.500	.778	668	21.5	17.0	34.5	3025
7	42	4.500 x 3.500	.778	668	23.0	Out	33.8	3025
7	42	4.125 x 3.500	.848	565	21.8	17	33.0	2560

Appendix B—Tornado 14 Detailed Description

(a) Manufacturer	
Name	Wright Aeronautical Corporation
Location of factory	Paterson, New Jersey
(b) Model and Rating of Engine	
Model	Tornado – 14 (T-14)
Normal Rated Power	500 HP at 3800 RPM
Military Rated Power	566 HP at 4150 RPM
Take-off Power	600 HP at 4150 RPM
(c) General Form	
Bore	4.250"
Stroke	3.625"
Number of Cylinders	14
Total piston displacement	720 cu in
Piston arrangement	2 rows in-line, 7 blocks radial
Cooling Medium	Liquid (Ethylene Glycol)
Compression Volume Ratio	7.2:1
(d) Crankshaft	
Material	Steel – AMS 6252
Journal Diameter	
Front Main	2.435"
Center Main	5.810"
Rear Main	3.497"
Crankpin Diameter	2.125"
Number of main bearings	3
Number of crank throws	2
Counterweight Construction	Integral with end main bearing, separate from crankpins
Number	2
Method of attachment	Clamp joint to the crankpin
Damper type	Weight suspended in the counterweight
(e) Propeller Hub	
Type	50 spline
(f) Connecting rods	
Length center to center	
Master rod	8.375"
Articulated rods	6.125"
Type	Solid
Material	Steel – AMS 6412
Bearing in rod	Solid
Weight	
Master rod	
Piston end	0.912 lbs.
Crankshaft end	4.765 lbs.
Articulated rod	
Piston end	0.530 lbs.
Crankshaft end	0.450 lbs.

(g) Bearings – Main	
Material	Copper lead, steel back – AMS 4820
Type	Split radially
(h) Piston	
Material	Aluminum alloy – USA Spec. 57-153-
Weight with rings, pin and pin retainers	2.80 lbs.
(i) Piston Pin	
Diameter	1.000
Retaining device	Plug – aluminum
Material	Steel – AMS 6250
(j) Cylinders	
Material	
Block	Beryllium Bronze W.A.C Spec. 6871
Sleeve	Steel – AMS 6250
(k) Valve Seats	
Material	Steel – AMS 5700
Form	Conical surface
Angle	45° 52'
(l) Valves	
Number per cylinder	2
Form	poppet
Material	
Intake	Steel – W.A.C. 8166
Exhaust	Steel – AMS 5700
Seat Angle	44° 22'
Lift	0..375"
Diameter	2.140"
Internal cooling (exhaust valve only)	Sodium and potassium
(m) Valve Springs	
Type	Helical
Number per valve	3
Load	
Valve open	189.6 lbs.
Valve closed	102.3 lbs.
Length	
Inner	
Free	1.810"
Valve closed	1.435"
Valve open	1.060"
Intermediate	
Free	1.910"
Valve closed	1.470"
Valve open	1.095"
Outer	
Free	1.960"
Valve closed	1.475"
Valve open	1.100"
Material	Steel Wire – USA 48-7 Grade WD-6150

(n) Valve Timing	
Inlet Opens	41° BTC
Inlet Closes	63° ABC
Exhaust Opens	74° BBC
Exhaust Opens	65° ATC
Provisions for adjusting	None
(o) Camshaft	
Type	Overhead
Number	7
Diameter	1.000"
Valve operating means	Rocker arm
Number of cams on shaft	4
Radius of heel of cam	0.587"
(p) Gun Control Drive	None
(q) Ignition	
Type	Dual
Timing	Intake — 30° BTC, Exhaust — 30° BTC
(r) Carburetor	
Number	1543
Vendor	Holley Carburetor Co.
Type	1375F
Altitude mixture control	None*
Settings	Experimental
Heater	No*
(s) Crankcase	
General form	Radial
Method of servicing crankshaft journals	Main bearings bolted in separate crankcase sections
(t) Coolant Pump	
Type	Centrifugal
Ratio of crankshaft speed to pump speed	4:3
Packing gland type	Ambest Metallic
(u) Oil Pump	
Type	Gear
Capacity at rate speed	150 lbs./min at 50 psi
(v) Blower	
Type	Centrifugal
Size	11" diameter
Ratio of speed to crankshaft speed	3.57:1 and 5:1
Number of vanes	22
(w) Fuel Pump	
Type	Vane
Speed	½ crankshaft
Capacity	360 gal/hr
(x) Generator	None*

*This engine is for test purposes only. It is not to be used as a flight engine.

Appendix C—Tornado 14 Tests

Account of the 50-Hour Development Test of the Wright Tornado 14[1]

July 23, 1940

Engine No. 11761 was started at 9:55 PM. After running for 10 minutes the engine was stopped to allow the replacement of a leaky Weatherhead hose line supplying oil to the nose. At 10:45 PM, the run-in was started and was continued until midnight.

July 24, 1940

At 12:50 AM, following 30 minutes running at 2,800 rpm, the engine was stopped to allow the installation of a booster pump in the oil-in line. The engine was stated for continuation of the run-in at 1:10 AM. The run-in was completed, a magneto check taken, and the engine shut down at 2:35 AM. The oil sumps were drained and a ¼" hexagonal nut was placed in back of the oil pressure relief valve spring to increase oil pressure. A 4:25 AM the engine was again started and five hours of endurance at 500 hp/3,800 rpm were run. Then, 10 five-minute take-off runs were made at 600 hp/4,150 rpm, each preceded by a 10-minute idle period. At 12:20 PM the engine was stopped and the sumps drained. General engine check were made before starting again at 3:30 PM to continue with an endurance run at 450 hp/3,670 rpm. At 7:30 AM the engine was stopped and a second booster pump installed in the oil-in line to decrease suction at the inlet to the engine pressure pump. At 10:05 PM the engine was started for continuation of the endurance. At midnight the total time was 17:30 hours.

July 25, 1940

The endurance at 450 hp/3,670 rpm continued until 10:35 when a stop was made for general engine checks. At 12:30 PM the endurance was continued. At 4:40 PM an oil leak made it necessary to stop. As chips of heavy metal were found in the front oil sump, the engine was removed from the stand. The total time on the engine was 32:15 hours.

September 20, 1940

The engine was again mounted on the test stand. After the start, the oil filter pressure outlet was found to be open. After connecting the line to the oil filter pressure gage, the engine was again started and a five-hour run-in was started at 8:30 PM. At midnight the total time on this installation of the engine was 3:40 hours.

September 21, 1940

A magneto check was taken at completion of the run-in. Running on the right magneto alone caused the engine to misfire. The exhaust spark plug leads to cylinders 9 and 10 were found crossed. Ignition wiring was corrected and the engine started at 7:05 AM. After a one-half hour run on non-lead fuel, the engine was shut down at 7:55AM for the weekend. Total time for this installation was 6:50 hours.

September 23, 1940

At 8:35 AM the engine was started and the previous endurance test at 450 hp/3,670 rpm was resumed. While taking the oil flow measurement at 11:30 AM, the oil in the supply tank was allowed to drop below the oil suction line and the oil pressure dropped suddenly. The engine was stopped and the sumps were drained. As there was no indication of bearing failure, the endurance test was continued. At the completion of ten hours of endurance, the engine was run for one-half hour on non-leaded fuel, which completed the fifty hours of endurance. The running time for this installation of the engine was 18:20 hours.

Account of the 150-Hour Endurance Test of the Wright Tornado 14[2]

1. The 150-hour endurance run of the type test followed the limits as prescribed in Specification No. AN 9502, item F-3d, but no attempt was made to make the run continuous throughout the 150 hours. Five complete and three partial disassembles were made during the test so the engine parts could be inspected.

2. The items contained in F-3d were run in the sequence listed. However, at the completion of the test it was noted that item F-3d-5 had be deviated from. Periods requiring 80 percent normal rated power and 93 percent normal rated speed (400 hp/3,535 rpm) were actually run at 80 percent normal rated power and 89 percent normal rated speed (400 hp/3,385 rpm). BMEP was necessarily higher, resulting in a more severe test than that specified.

3. All testing was done in the Experimental Test Laboratory of Wright Aeronautical Corporation, Paterson, New Jersey. Power was absorbed by a combination of water brake and electric dynamometer connected directly to the propeller shaft.

4. The following operating limits were observed throughout the test:

Oil inlet temperature	180–190°F
Layshaft oil pressure	50–55 psi
Carburetor air temperature	70–80°F
Fuel temperature	60–70°f
Coolant temperature	245–255°F

Coolant flow sufficient to produce a nominal 20° rise

5. Two external booster pumps supplied oil pressure because the standard C14 engine oil pump would not supply the necessary pressure. Oil was admitted to the engine pump inlet and the pump idler gear was removed to facilitate the flow. Pressure was manually controlled by bleeding the "pressure out of pump" line

6. Coolant pressure was supplied by the engine coolant pump and the flow was regulated by a gate valve in the "out of engine" line.

7. Exhaust gasses were collected through a water-cooled exhaust ring, which was connected to engine 90° stub stacks by loose fitting sleeves. An exhaust blower prevented back pressure in the system.

8. Fuel flow was measured by weighing a predetermined amount of fuel on a scale. The scale beam, through electrical switches, started and stopped an rpm counter which measured the exact number of engine revolutions during a period. All reading periods were in excess of one minute.

9. Readings were taken every 15 minutes through the test except during the "take-off" cycles when one reading was taken during each five-minute take-off period and another during each 10-minute "idle" period.

March 1, 1941

The engine was installed on X-10 dynamometer stand and run-in for five hours. Prior to this installation, the engine had been rebuilt seven times and had accumulated 157:25 hours running time.

March 3-4, 1941

The engine was taken to the Experimental Assembly Department and the following units were disassembled for inspection:

1. Crankcase power section
2. Cylinder blocks
3. Front assembly drive section

The nose section and front and rear supercharger housings were not disassembled. Inspection of the parts revealed:

1. The front master rod bearing had surface cracks at the front end around No. 6 knuckle pin position. Total running time on this bearing was 139:50 hours. The cracking was attributed to non-uniform copper and lead distribution within the bearing material. The bearing was replaced.

2. No. 5 and No. 9 intake valves showed magnetic indications and were replaced. Later investigation of numerous valve rejections revealed faulty magnetic inspection technique was causing indications on good valves.

3. The top ring of No. 7 piston had a short piece broken off one end.

4. The No. 4 cam drive idler gear bushing had backed out of position 0.007" due to insufficient shrink fit. A new one was installed with the proper shrink fit and no further trouble was encountered.

5. The No. 4 magneto drive gear bushing was scratched and slightly heat discolored on its thrust face. Inspection of the other bushings showed that high loading on a small portion of the thrust face was casing this condition. Removal of the high spots corrected this.

March 10-14, 1941

Engine assembly was completed and the engine was installed on the X-10 stand for continuation of the endurance test. This portion of the test was to consist of item F-3d-2 with the take-off limit at 600 hp/4,150 rpm.

During the five-hour run-in, a cylinder hold-down stud at the rear of No. 7-8 cylinder block broke. It was replaced while the engine remained on the test stand. At the completion of the run-in, a take-off power checkpoint was attempted, but the engine backfired and the power dropped. Investigation revealed that No. 13 intake valve had stretched and was remaining open. It was also noted during the run-in that oil flow was slightly higher than on the previous test. This condition was also investigated during the shut down. The front main oil seal ring was found crimped out of position. Replacement of this part required a partial disassembly of the crankcase.

March 14-24, 1941

The engine was taken to the Experimental Assembly Department and disassembled sufficiently to replace the main bearing oil seals. Upon disassembly, both front and rear main bearings were discovered "picked out". While this failure is attributed to non-uniformity of the bearing material, it is possible that low oil pressure in this region due to the seal, which had crimped out of position, could have initiated the picking out. It is interesting to note that no further main bearing trouble was encountered after the oil seals were replaced. They were held in position during assembly by an adhesive compound.

Three exhaust valves had slightly collapsed heads due to overheating during the run-in, and three intake valves had lost their hardness due to the same reason. Overheating was caused by incorrect fuel flow. The three intake valves were replaced and a complete new set of exhaust valves was installed as a precautionary measure

March 24-25, 1941

Engine assembly was completed and the engine was installed on X-10 test stand for continuation of the test. After a five-hour run-in and 20 take-off cycles were run according to item F-3d-2 (600 hp/4,150 rpm take-off limit) the engine backfired and lost speed due to a broken articulated rod.

March 25, 1941 – April 9, 1941

The engine was removed to the Experimental Assembly Department for complete disassembly and inspection. The inspection revealed:

1. No. 14 articulated rod broke at the highly stressed area near the piston pin hole and damaged three cylinder sleeves, one exhaust valve and one high-speed power shaft. These parts were replaced. The articulated rods were known to be structurally weak and were being used only until a redesigned and stronger set of rods could be procured. These new rods, along with the necessary pistons, piston rings and piston pins were introduced at this time, and no more articulated rod trouble was encountered.

2. The No. 1 magneto drive gear and bushing had seized because of poor alignment. New parts were installed and no further trouble was encountered.

3. One of the exhaust valve seat inserts was loose in the cylinder block and a complete block assembly had to be installed because an oversize insert was not available.

4. A new reduction gear pinion carrier was also installed at this time. The used part was still in good condition, but a rework, which improved the oil flow characteristics, was available and was included.

April 9-18, 1941

After completion of assembly, the engine was mounted on X-10 test stand for further testing. After a five-hour run-in, 60 take-off cycles were run in accordance with Item F-3d-2. The take-off limit was 600 hp/4,150 rpm. Stops were made to do the following servicing:

1. After the 14th take-off cycle, it was necessary to replace the ignition harness. At the same time all spark plugs were replaced. Again after the 36th take-off, the ignition harness was replaced. Both replacements were necessary because overheating had burned the insulation on some of the leads. This condition might well have been expected since the X-10 stand was not equipped with a cooling air blower and the harness could not be subjected to an air blast. After the 48th take-off, two spark plug cigarettes were replaced because they were burned.

2. After the 49th take-off, the carburetor compensator unit was repaired because of excessive fuel flow.

3. At the completion of the 56th take-off, two rear cylinder hold-down studs were found broken. They were replaced while the engine remained on the test stand.

4. While attempting the 57th take-off, ignition harness trouble was again encountered, and since only one hour of the test remained, jumper ignition leads were installed to four spark plugs.

The test run was then completed without further delay.

The only delay that could be attributed to an engine part during this installation was caused by breakage of the hold-down studs. The engine design permits these studs to be strengthened either by a design change or size increase.

April 18-29, 1941

The engine was removed to the Experimental Assembly Department and completely disassembled and inspected. The inspection revealed the following:

1. The front master rod bearing was picked out due to non-uniformity of the bearing material. It was replaced.

2. Three exhaust valve heads were slightly collapsed. The valves were replaced.

3. The intermediate power shaft gear was pitted at the pitch circle. Analysis of this part showed that it did not have proper tooth contact with its mating gear, and that the

hardness of its teeth was below required limits. Even under these adverse conditions, the gear had accumulated 145:00 hours running time. No further trouble was encountered with this part once it was replaced.

4. An exhaust cam roller and a high-speed power shaft pinion were slightly picked out. Both parts were still usable, but were replaced as routine practice. A new rocker and shaft assembly was included with the cam roller replacement.

During this rebuild, the carburetor was completely overhauled. The ignition harness was also reworked to provide air cooling in an effort to prevent the breakdown of ignition cables and terminals.

April 29 – May 7, 1941

Engine assembly was completed and the engine was installed on the X-10 test stand for continuation of the test. This portion consisted of item F-3d-3 with test limits of 566 hp/4,150 rpm and 124 hp/2,500 rpm.

After a five-hour run-in, the 60 cycles required by this element of the specification were completed without delay due to the malfunctioning of any engine part.

May 7 – 22, 1941

The engine was taken to the Experimental Assembly Department and completely disassembled and inspected. The inspection revealed:

1. It was necessary to replace the intermediate power shaft bushing because it had cracked circumferentially about half the distance around and approximately midway between its oil annulus and ids flanged portion. This bushing was made from cast material, and the crack was a result of poor material structure. The bushing was originally full floating. When replaced, it was installed with a shrink fit in the front crank cheek journal.

2. One of the exhaust valve seat inserts backed out of position. A new insert with the proper shrink fit was installed.

3. The top ring of two pistons broke and mutilated the piston ring grooves. New pistons and rings were installed.

4. One of the three rear end drive gear oil seal rings was broken. This was an isolated case.

5. The hydro oil seal support backed out of position slightly. It was pushed back into place and reworked to provide for a locking pin. This was a standard C-14A part and the rework was standard for this item.

6. Two intake valves showed magnetic indications and were replaced. As mentioned before, faulty inspection technique may have caused these indications.

May 22 – 28, 1941

Engine assembly was completed and the engine mounted on the X-10 test stand to continue with Item F-3d-5. Test limits should have consisted of alternate periods at 500 hp/4,150 rpm and 400 hp/3,535 rpm, but at the completion of the test it was noted that the actual limits were 500 hp/4,150 rpm and 400 hp/3,385 rpm. This changes the speed limit from 93% normal rated speed to 89% normal rated speed at the lower power setting. After a five-hour run-in, 12:15 hours of endurance testing were completed before two rear cylinder hold-down studs broke and necessitated a shut down. When the spark plugs were removed from the offending cylinder block, a spark plug insert was found loose. A new insert was installed, the studs replaced, and the remaining portion of the 25 hour run was completed without delay.

May 28 – June 14, 1941

The engine was removed to the Experimental Assembly Department and the cylinder blocks were removed and disassembled for inspection.

The engine was found satisfactory in every respect except for threads in two spark plug holes. New inserts were installed in these positions. No other new parts were added.

June 14 – 20, 1941

After completion of engine assembly, the engine was mounted on the X-10 test stand for running of item F-3d-6 with limits of 500 hp/3,800 rpm and 350 hp/3,385 rpm.

After a 2:30 hour run-in, 17:45 hours of endurance were completed when the engine ran out of fuel. Upon restarting, the engine backfired and investigation revealed that one of the rocker supports was broken from fatigue. It was replaced along with a new support bolt and a new camshaft bushing. These two parts were damaged when the support broke. The remaining portion of the 25-hour run was completed without further engine trouble.

June 20 – 23, 1941

After a routine service without removal from the test stand, the engine ran an additional 25-hour test of item F-3d-7 consisting of alternate periods at 500 hp/3,800 rpm and 300 hp/3,195 rpm. It was necessary to stop after 16:30 hours of this run to replace a rear cylinder hold-down stud. The remainder of the time was then completed without delay.

June 23 – July 11, 1941

The engine was taken to the Experimental Assembly Department and completely disassembled and inspected. The inspection revealed:

1. Three high-speed power shaft pinion spacers had steel particles imbedded in them. They were otherwise in good condition and were replaces to protect their mating.parts.

2. The front and rear master rod bearings had fine surface cracks, and the rear one was also picked out. Non-uniform material structure caused both failures.

3. One articulated rod knuckle pin bushing was cracked at both ends where the bushing material had been spun over the inner corners of the chamfers at both ends of the pin hole. The material had evidently been fatigued or cracked during the spinning operation during assembly. The knuckle pin and articulated rod were in excellent condition.

4. Four damper pin bushings in the front crank cheek were worn and corroded. All were still usable but were replaced as routine practice.

July 11 – 14, 1941

Assembly of the engine for the last portions of the test was completed and the engine was mounted on X-10 test stand

.

After a five-hour run-in, item F-3d-8 was completed without delay. Test Limits were 300 hp/2,660 rpm.

One hour of running at 500 hp/4,180 rpm in accordance with item F-3d-9 was completed when the engine power output became unstable and then cut out. Adjusting the left hand magneto breaker points and installing a new set of points in the right hand magneto remedied this difficulty, and the engine completed the 150-hour test without further delay

July 14, 1941

The engine was taken to the Experimental Assembly Department and completely disassembled and inspected. The inspection revealed:

1. The front and rear master rod bearings were cracked and as in all other instances, the cracking was the result of non-uniformity of the bearing material;

2. OD of one rocker roller was pitted.

All other engine parts were in good condition for further running. Clearance checks were made and no part was found to be worn excessively.

---

[1] Swords, *Report on 50-Hour Development Test.*

[2] Lundquist, *Summary of150-Hour Endurance Test.*

Appendix D—Reprints of Tornado 42 Specifications

WRIGHT AERONAUTICAL CORPORATION

PATERSON, NEW JERSEY, U.S.A.

Date: January 4, 1939
Revised: May 23 1939
Revised: July 11, 1939

PRELIMINARY
W.A.C ENGINE SPECIFICATION NO. 617-B
WRIGHT TORNADO ENGINE
MODEL ______ GEARED

FOR
U. S. ARMY AIR CORPS OR
U. S. NAVY BUREAU OF AERONAUTICS
CONSTRUCTED IN ACCORDANCE WITH
SPECIFICATION NO. AN-9500
DATED MARCH 1, 1939

(W.A.C. Model Designation No. Tornado 617)

MODEL SPECIFICATION FOR
ENGINE, AIRCRAFT
GEARED WRIGHT TORNADO 617
FOR USE WITH EXHAUST TURBO SUPERCHARGER

APPLICABLE SPECIFICATIONS

A-1. The current issue of the following specifications, in effect on the date of issuance of engine bids, shall form a part of this specification:

AN-9500 Engine, Aircraft, General Specifications.

B. TYPE

B-1. This specification covers the requirements for the Tornado 617 engine.

C. GENERAL REQUIREMENTS

C-1. The requirements of specifications listed in Section A shall constitute general requirements for the engine.

D. DETAIL REQUIREMENTS

D-2. Drawings. – The following Wright Aeronautical Corporation drawings form part of this specification:

Dwg. No. **Drawing Title**
413460 – Engine Assembly, Basic (showing accessory drive oil seals)
413465 – Installation Drawing (showing clearances for engine accessories and their removal)
413461 – Priming system assembly
412462 – Screen Assembly, Air Inlet
110559 – Gasket, Screen, Air Inlet
110560N – Carburetor, Stromberg
110323, 64174, 68429, or 69561 – Spark Plug Assembly
61722 & 61800 – Terminal, Spark Plug
413465 – Installation, Cylinder Cooling Baffles, Lubrication System Diagram, Clearance Chart
110561N – Ignition Assembly, Radio Shielded
84864 – Tool Kit Assembly

D-3. Acceptance.- Acceptance of this engine shall be predicated on the satisfactory completion of a type test and acceptance test conducted in accordance with Specifications AN-9502 and AN-9503 respectively, with exceptions as specified in paragraphs D-3a and D-3b herein.

D-3a. Type Test (Reference AN-9502, F-3c)

Power calibration of the engine shall be made with only the magneto, tachometer, and fuel pump drive loaded.

D-3b. Carburetor metering runs shall be conducted in accordance with Spec. AN-9503, paragraphs F-5d to F-5d (7) inclusive.

D-4. Weight.- The dry weight of the basic engine complete without exhaust turbo supercharger and without intercoolers, shall not exceed 2400 pounds as follows:

D-4a. The basic engine shall include integral supercharger, supercharger drive mechanism, propeller reduction gears, coolant pumps and piping on the engine, engine lubrication system oil pumps, starter connection, including starter dog, gun synchronizing impulse generator drives (but not impulse generators), tachometer drives, fuel pump drive, generator drive, vacuum pump drives, propeller governor drive, all piping and controls between engine parts, carburetor and adapter or equivalent, carburetor screens and gaskets, magnetos, ignition cable assembly, spark plugs, priming system on engine, exhaust flanges and gaskets, cooling air deflectors and baffles and accessory drive covers.

D-5. Performance Characteristics.- The engine shall be rated as follows, using fuel conforming to Specification AN-9531 and oil conforming to Specification AN-9532, Grades 1120 and an exhaust turbo supercharger of suitable capacity. Refer to Note H-5 of this specification.

The terms used herein and the standard conditions are in accordance with the applicable definitions specified in Specification AN-9503.

2000 BHP at 845 Propeller RPM at Sea Level.

2000 BHP at 845 Propeller RPM at 25000 feet, normal rated altitude.

2350 BHP at 925 Propeller RPM take-off for three minutes.

2350 BHP at 925 Propeller RPM military rating at 25000 feet, military rated altitude

1015 Propeller RPM, rated overspeed RPM

D-5(1). When necessary, demonstration of the guaranteed performance shall be made on the dynamometer under simulated altitude conditions and with best power mixture except that take-off power shall be demonstrated with the take-off specific fuel consumption specified on page 9 of this specification. During such demonstration standard pressure shall be maintained at the fitting in the top flange of the carburetor or at the pressure indicating fitting provided at the entrance to the engine induction system.

D-5a. The curves shown on page 9 shall form a part of this specification, and together with the ratings given above shall constitute the power and specific fuel consumption guarantees. Data called for in Specification AN-9501, paragraphs D-5a(5) to D-5b(8) inclusive shall be furnished one month after delivery of the first engine, with exceptions as listed in paragraph D-5b herein.

D-5b. Performance Curves (Ref. Spec AN9501, Par. D-5b)

D-5b(1). These data shall be plotted against BMEP and shall cover the range from 30% to 70% rated power and 50% rated speed to rated speed. The data above 70% rated power shall be plotted on propeller load through rated power and speed and shall include a point at take-off power and speed.

D-5b(2). This curve shall not be furnished.

D-5b(3). These data shall cover the range defined in paragraph D-5b(1).

D-5b(5). These data shall be plotted against BMEP and shall cover the range from 70% rated power to take-off power and from 85% rated speed to take-off speed. The range shall be further defined y the load compensation characteristics of the carburetor.

D-5b(6). These data shall cover the range defined in paragraph D-5b(5).

D-5b(7). These data shall be plotted against BMEP and shall cover the range from 30% to 70% power and from 50% rated speed to maximum cruising BMEP. The range shall be further defined by the load compensation characteristics of the carburetor

D-5c. Specific Oil Consumption.- The specific oil consumption shall not exceed 0.035 lb./BHP/Hr. at normal rated power and speed and 0.025 lb./BHP/Hr. at 70% normal rated power and 89% normal rated speed.

D-6. The complete engine shall function satisfactorily up to the service ceiling of the airplane provided the required fuel pressure and adequate cooling air are supplied to the engine.

D-7. Propeller.- The engine shall have a No. 60 propeller shaft end. Provision shall be made for a hydro-matic type of propeller control mechanism.

D-7a. Ends, Nose & Propeller Shaft (Ref. Spec. AN-9506, Par. E-1b(1), Fig. 3). The cutter used for removing the incomplete thread shall have a minimum diameter of 2-3/8".

D-7b. Drive & Hydro Control System, Prop. Governor (Ref. Spec. AN-9507, Fig. 1). The chordal space of the spline shall be 0.0791 plus or minus 0.001".

D-12. Overall Dimensions.- The overall dimensions of the engine shall not exceed the following:

Length	93 inches
Diameter	37 inches

D-14. Preparation for Storage.- The engine shall e prepared for storage prior to shipment, in accordance with Army Technical Order No. 02-1-1, using compound 3568-B. Prior to the run with unleaded fuel, the oil shall be changed to new (unused) oil.

D-15. Protective Coatings and Finish.- Protective coatings and finish shall be in accordance with the manufacturer's drawings.

D-16b. Parts List of the Engine.- The parts list of the engine which successfully completes Government tests shall constitute a requirement of this specification.

D-18. Propeller Drive.- The engine shall be equipped with a reduction gear. The direction of propeller rotation, when viewed from the anti-propeller end, shall be clockwise.

D-19. Impeller Gear.- The impeller gear ration shall be single stage one speed and the impeller shall be 12 inches in diameter.

D-20. Pistons.- The engine shall be fitted with pistons of 7.0:1 compression ratio.

D-22. Crankshaft Torsional Vibration (Ref. Spec. AN-9504, Par. E-4). Crankshaft torsional vibration characteristics will be treated according to paragraph E-4.

D-23a(1). Spark Plugs.- The engine shall be fitted with BG298GS, Aero LS1AC, LS38C or Bendix 5S5 spark plugs, furnished in accordance with the Manufacturer's drawings, having 1" hexagon flanges and 9/16"-27 threads at the terminal ends.

D-23a(1)a. (Ref. Spec AN-9509, Par. F-11). Spark plugs may be selected at periodic intervals and subjected to the other tests specified in Specification AN-9509.

D-23a(1)b. (Ref. Spec AN-9509, Par. F-2b(8).). The engine manufacturer will not be responsible for the performance of the engine unless the tests of paragraph F-2b(8) and F-2b(9) of Specification AN-9509 are run during spark plug type tests.

D-23a(1)c. (Ref. Spec AN-9509, Paragraph G-1a). Corrosion preventive compound shall be in accordance with manufacturer's specification.

D-23b. Radio shielded Ignition Assemblies.- The engine shall be equipped with radio shielded ignition assemblies.

D-23b(1). (Ref. Spec AN-9510, Paragraph D-1a(1).). The minimum radius of the spark plug elbow3s shall be 5/16" approximate.

D-23b(2). Ref. Spec AN-9510, Paragraph E-1a). The threaded connection to the spark plug terminal shall be in accordance with manufacturer's drawings.

D-23b(3). Ref. Spec AN-9510, Paragraph E-6). The requirements of reference paragraph shall not be applicable.

D-23d. Magnetos.- the engine shall be equipped with Scintilla magnetos in accordance with manufacturer's drawings.

D-23g. Cooling.- Provisions will be made for cooling the interconnecting parts of the ignition system.

D-24g. Oil Cleaner.- Foreign matter being removed by the cleaner shall not re-enter the lubricating system under normal operating conditions.

D-24i. Temperature Measurements.- Clearance within the engine shall be provided for the sensitive portion of the bulb which shall not exceed 17/32" diameter.

D-24j(1). Oil Connections.- Oil inlet elbows shall no be furnished with the engine.

D-24l. Accessory Oil Return.- Two 3/8" pipe threaded openings will be provided in the rear section of the engine but the contractor will not be responsible for the engine operation with the vacuum pump oil separator discharge connected to these openings.

D-24q. Crankcase Breather.- The breather shall terminate in a hose nipple connected in accordance with the manufacturer's drawings. The opening will normally be covered by a breather cap. The contractor assumes the responsibility only for the breathing capacity adequate to handle an additional two cubic feet of air per minute and not the discharge from the oil separator.

D-25. Fuel Metering System.- The engine shall be equipped with a Stromberg type carburetor system.

D-25b(2). Carburetor (Navy).- The carburetor setting determination shall be in accordance with A.C. Spec. No 28155-B.

D-26a. Fuel Priming System (Navy).- Provision shall be made for priming the engine.

D-32. Exhaust System.- The engine shall not be equipped with a hot spot.

D-32a&b. Exhaust Flanges. (Army and Navy).- Exhaust flanges and gaskets shall be furnished with the engine.

D-36. Accessory Drives.- The speed of each accessory drive at normal rated engine speed, the maximum permissible torque in inch pounds for continuous operation, the maximum permissible static torque in inch pounds and the direction of rotation when looking at the end of he accessory drive shaft in the engine shall be as follows:

Accessory Drive	**Normal RPM**	**Maximum Continuous Torque (Inch Lbs)**	**Maximum Static Torque (Inch Lbs)**	**Rotation**
Magneto				CW
Gun Synchronizer	845			L-CW, R-CCW
Hydraulic Pump	3040			CW
Fuel Pump	1900			CCW
Vacuum Pump	3040			CW
Generator	3040			CW
Tachometers	1900			L-CW, R-CCW
Propeller Governor	2280			CW

D-36a. Starter.- The starter mounting pad diameter shall be seven inches. The direction of rotation when looking at the starter dog attached to the engine shall be anticlockwise, and the number of jaws on the starter dog shall be three. The starter dog shall be detachable. The clearance provided for the starter shall be in accordance with the contractor's installation drawing.

D-36b. Generator.- A standard 6 inch diameter, 6 stud mounting pad with involute spline drive will be supplied with on the engine.

D-36c. Power Take-off Drive.- The generator drive will be suitable for use as a Type II gear box drive.

D-36d. Fuel Pump.- A standard square 4 stud mounting pad with involute spline drive will be furnished on the engine, in accordance with Specification AN-9519, Figure 1.

D-36e. Gun Synchronizer Impulse Generator.- Two Type I gun synchronizer impulse generator drives in accordance with Specification AN-9520 will be provided on the engine.

D-36f. Vacuum and Hydraulic Mechanism Oil Pumps.- Two large Type II, vacuum pump mounts and .600 P.D. spline drives without tongue and groove coupling will be furnished on the engine. The chordal space of the spline shall be .0791” plus or minus .001”.

D-36g. Tachometer Drives.- One each of Type I and Type II mounting pads and drives will be provided on the engine. Although clearance will be provided for an electric tachometer, the contractor will not assume responsibility for the tachometer drive assembly when the electric tachometer generator is mounted directly on the threaded outlet.

D-42. Accessories, Tools, Etc. (Navy).- The following items shall be furnished with the engine:

(a) First Installation Engine. (Experimental)

Blank covers for all accessory drives.

Set of engine tools for maintenance purposes (Tool Kit).

Engine instruction book.

(b) Service Engine.- Not Applicable.

F-2c(2) Cooling Liquid.- The cooling liquid outlet temperature shall be 121°C. (250°F.).

H. NOTES

H-1 (Navy) Classification of Engines.- The engine specified above is classified as an “experimental” engine.

H-2. Not applicable.

H-3. Not applicable.

H-4. Not applicable

H-5 The engine performance specified in paragraph B-5 of this specification is predicated on the use of a fuel having distillation characteristics as follows:

10% point 156°F. Maximum

50% point 212°F. Maximum

90% point 257°F. Maximum

These distillation characteristics are in substantial agreement with those of the fuels in current production.

H-6. The engine manufacturer reserves the right to change the rated speed, compression ratio, carburetor, supercharger impeller diameter and gear ratio, spark plugs, and other parts of the design which affect volumetric efficiency and/or combustion efficiency, in order to meet the guaranteed engine performance, provided the airplane performance is not affected. These changes will be incorporated I the type test engine and will be made a definite part of the final coordinated specification prior to release by the Materiel Division.

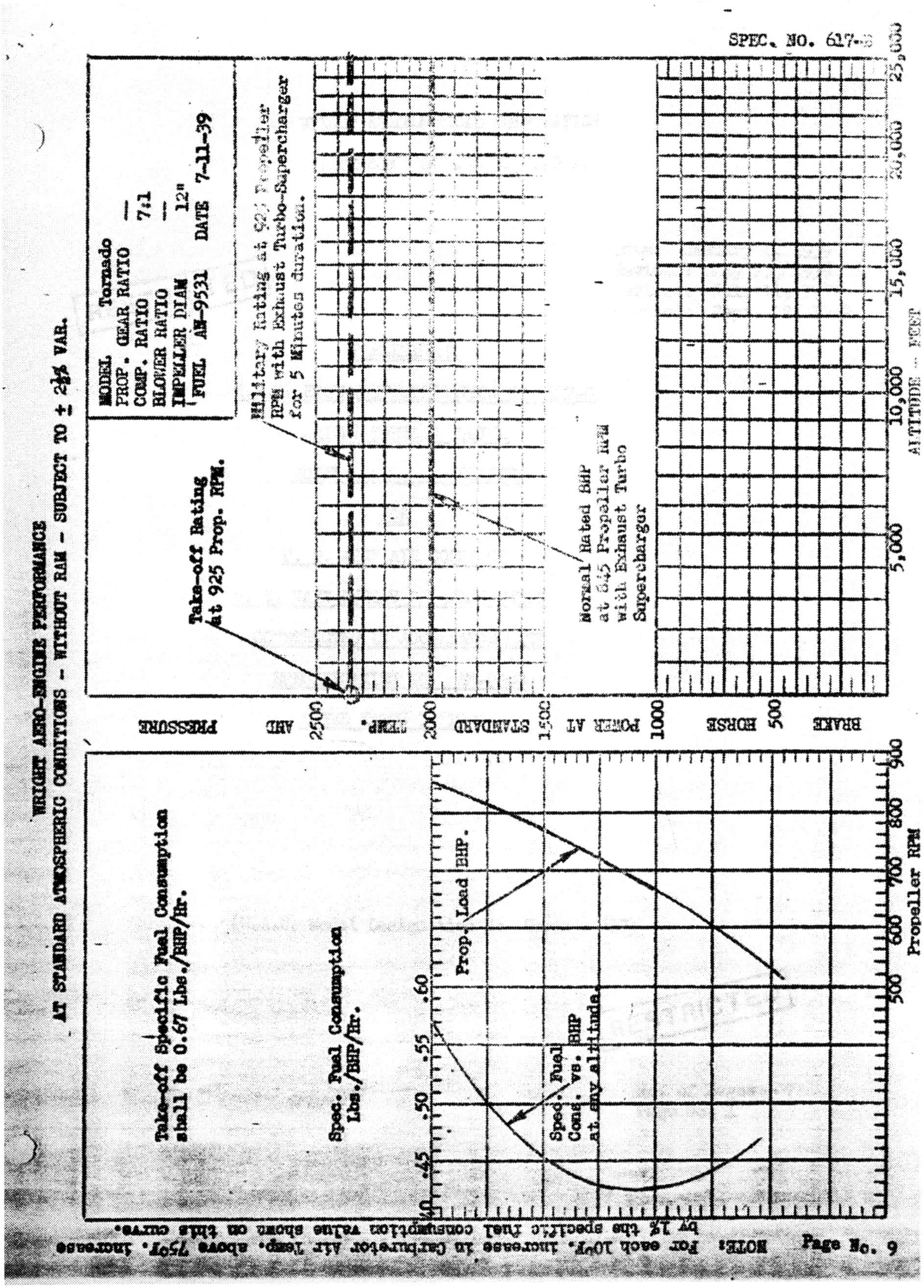
WRIGHT AERO-ENGINE PERFORMANCE
AT STANDARD ATMOSPHERIC CONDITIONS - WITHOUT RAM - SUBJECT TO ± 2½% VAR.
Take-off Specific Fuel Consumption shall be 0.67 Lbs./BHP/Hr.
Spec. Fuel Consumption Lbs./BHP/Hr.
.45 .50 .55 .60
Prop. Load BHP.
Spec. Fuel Cons. vs. BHP at any altitude.
500 600 700 800 900
Propeller RPM
BRAKE HORSE POWER AT STANDARD TEMP. AND PRESSURE
500 1000 1500 2000 2500
Take-off Rating at 925 Prop. RPM.
MODEL Tornado
PROP. GEAR RATIO —
COMP. RATIO 7:1
BLOWER RATIO —
IMPELLER DIAM 12"
FUEL AN-9531 DATE 7-11-39
Normal Rated BHP at 845 Propeller RPM with Exhaust Turbo Supercharger
5,000 10,000 15,000 20,000 25,000
ALTITUDE - FEET
NOTE: For each 10°F. increase in Carburetor Air Temp. above 75°F. increase by 1% the specific fuel consumption value shown on this curve.
Page No. 9

WRIGHT AERONAUTICAL CORPORATION

PATTERSON, NEW JERSEY, U.S.A.

Date: January 4, 1939
Revised: May 23 1939
Revised: July 11, 1939
Revised: Sept. 28, 1939

PRELIMINARY
W.A.C ENGINE SPECIFICATION NO. 617-C
WRIGHT TORNADO ENGINE
MODEL GEARED

FOR
U. S. ARMY AIR CORPS OR
U. S. NAVY BUREAU OF AERONAUTICS
CONSTRUCTED IN ACCORDANCE WITH
SPECIFICATION NO. AN-9500
DATED MARCH 1, 1939

(W.A.C. Model Designation No. Tornado 617)

MODEL SPECIFICATION FOR
ENGINE, AIRCRAFT
GEARED WRIGHT TORNADO 617
FOR USE WITH EXHAUST TURBO SUPERCHARGER

A. APPLICABLE SPECIFICATIONS

A-1. The current issue of the following specifications, in effect on the date of issuance of engine circular proposals, shall form a part of this specification:

AN-9500 Engine, Aircraft, General Specifications.

B. TYPE

B-1. This specification covers the requirements for the Tornado 617 engine.

C. GENERAL REQUIREMENTS

C-1. The requirements of specifications listed in Section A shall constitute general requirements for the engine.

D. DETAIL REQUIREMENTS

D-2. Drawings. – The following Wright Aeronautical Corporation drawings form part of this specification:

Dwg. No.	Drawing Title
413460	– Engine Assembly, Basic (showing accessory drive oil seals)
413465	– Installation Drawing (showing clearances for engine accessories and their removal)
413461	– Priming system assembly
412462	– Screen Assembly, Air Inlet
110559	– Gasket, Screen, Air Inlet
110560N	– Carburetor, Stromberg, Spark Plug Assembly, Terminal, Spark Plug
413465	– Installation, Cylinder Cooling Baffles, *Lubrication System Diagram, *Clearance Chart
110561N	– Ignition Assembly, Radio Shielded

* Included in Instruction Book

D-3. Acceptance.- Acceptance of this engine shall be predicated on the satisfactory completion of an acceptance test conducted in accordance with Specifications AN-9503.

D-4. Weight.- The dry weight of the basic engine complete without exhaust turbo supercharger and without intercoolers, shall not exceed 2400 pounds as follows:

D-4a. The basic engine shall include integral supercharger, supercharger drive mechanism, propeller reduction gears, coolant pumps and piping on the engine, engine lubrication system oil pumps, starter connection, including starter dog, gun synchronizing impulse generator drives (but not impulse generators), tachometer drives, fuel pump drive, generator drive, vacuum pump drives, propeller governor drive, all piping and controls between engine parts, carburetor and adapter or equivalent, carburetor screens and gaskets, magnetos, ignition cable assembly, spark plugs, priming system on engine, exhaust flanges and gaskets, cooling air deflectors and baffles and accessory drive covers.

D-5. Performance Characteristics.- The engine shall be rated as follows, using fuel conforming to Specification AN-9531 and oil conforming to Specification AN-9532, Grades 1120 and an exhaust turbo supercharger of suitable capacity.

The terms used herein and the standard conditions are in accordance with the applicable definitions specified in Specification AN-9503.

2000 BHP at 845 Propeller RPM at Sea Level.

2000 BHP at 845 Propeller RPM at 25000 feet, normal rated altitude.

2350 BHP at 925 Propeller RPM take-off for five minutes.

2350 BHP at 925 Propeller RPM military rating at 25000 feet, military rated altitude

1015 Propeller RPM, rated overspeed RPM

D-5(1). When necessary, demonstration of the guaranteed performance shall be made on the dynamometer under simulated altitude conditions and with best power mixture except that take-off power shall be demonstrated with the take-off specific fuel consumption specified on page 7 of this specification. During such demonstration standard pressure shall be maintained at the fitting in the top flange of the carburetor or at the pressure indicating fitting provided at the entrance to the engine induction system.

D-5a. The curves shown on page 7 shall form a part of this specification, and together with the ratings given above shall constitute the power and specific fuel consumption guarantees. Data called for in Specification AN-9501, paragraphs D-5a(5) to D-5b(8) inclusive shall not be furnished.

D-5b. Performance Curves (Ref. Spec AN9501, Par. D-5b)

D-5c. Specific Oil Consumption.- The specific oil consumption shall not exceed 0.035 lb./BHP/Hr. at normal rated power and speed and 0.025 lb./BHP/Hr. at 70% normal rated power and 89% normal rated speed.

D-5d(4). Heat rejection data will be furnished after completion of model test.

D-7. Propeller.- The engine shall have a No. 60 propeller shaft end. Provision shall be made for a hydromatic type of propeller control mechanism.

D-7a. Ends, Nose & Propeller Shaft (Ref. Spec. AN-9506, Par. E-1b(1), Fig. 3). The cutter used for removing the incomplete thread shall have a minimum diameter of 2-3/8".

D-7b. Drive & Hydro Control System, Prop. Governor (Ref. Spec. AN-9507, Fig. 1). The chordal space of the spline shall be 0.0791 plus or minus 0.001".

D-12. Overall Dimensions.- The overall dimensions of the engine shall not exceed the following:

Length 95.5 inches

Diameter 37 inches

D-14. Preparation for Storage.- The engine shall e prepared for storage prior to shipment, in accordance with Army Technical Order No. 02-1-1, using compound 3568-B. Prior to the run with unleaded fuel, the oil shall be changed to new (unused) oil.

D-15. Protective Coatings and Finish.

D-15a(5).Magnesium alloy parts shall be anodized in accordance with W.A.C. Spec. 5742C.

D-15b(2). Exposed cadmium plate and stainless steel shall not be plated.

D-16b. Parts List of the Engine.- The parts list of the engine which successfully completes Government tests shall constitute a requirement of this specification.

D-18. Propeller Drive.- The engine shall be equipped with a reduction gear. The direction of propeller rotation, when viewed from the anti-propeller end, shall be clockwise. The propeller RPM at rated speed shall be 845.

D-19. Impeller Gear.- The impeller gear ration shall be single stage one speed and the impeller shall be 12 inches in diameter.

D-20. Pistons.- The engine shall be fitted with pistons of 7.0:1 compression ratio.

D-22. Crankshaft Torsional Vibration (Ref. Spec. AN-9504, Par. E-4). Crankshaft torsional vibration characteristics will be treated according to paragraph E-4.

D-23a(1). Spark Plugs.- The spark plug designations shall be provided prior to submission of the engine to acceptance test and deviations to Spec. AN-9509 shall be inserted.

D-23a(1)c. (Ref. Spec AN-9509, Paragraph G-1a). Corrosion preventive compound shall be in accordance with manufacturer's specification.

D-23b. Radio shielded Ignition Assemblies.- The engine shall be equipped with shielded ignition assemblies, in accordance with the contractor's drawings.

D-23b(1). (Ref. Spec AN-9510, Paragraph D-1a(1)

D-23d. Magnetos.- the engine shall be equipped with Scintilla magnetos in accordance with manufacturer's drawings.

D-23g. Cooling.- Provisions will be made for cooling the interconnecting parts of the ignition system. Temperature requirements will be met provided adequate cooling is supplied.

D-24g. Oil Cleaner.- Foreign matter being removed by the cleaner shall not re-enter the lubricating system under normal operating conditions.

D-24i. Temperature Measurements.- Clearance within the engine shall be provided for the sensitive portion of the bulb which shall not exceed 17/32" diameter.

D-24j(1). Oil Connections.- Oil inlet elbows shall no be furnished with the engine.

D-24l. Accessory Oil Return.- Two 3/8" pipe threaded openings will be provided in the rear section of the engine.

D-24q. Crankcase Breather.- The breather shall terminate in a hose nipple connected in accordance with the manufacturer's drawings. The opening will normally be covered by a breather cap.

D-25. Fuel Metering System.- The engine shall be equipped with a Stromberg type carburetor system. Deviations to Spec. AN-9515 will be inserted prior to the acceptance test of the engine.

D-32. Exhaust System.- The engine shall not be equipped with a hot spot.

D-32a&b. Exhaust Flanges. (Army and Navy).- Exhaust flanges and gaskets shall be furnished with the engine.

D-36. Accessory Drives.- The speed of each accessory drive at normal rated engine speed, the maximum permissible torque in inch pounds for continuous operation, the maximum permissible static torque in inch pounds and the direction of rotation when looking at the end of he accessory drive shaft in the engine shall be as follows:

Accessory Drive	Normal RPM	Maximum Continuous Torque (Inch Lbs)	Maximum Static Torque (Inch Lbs)	Rotation
Magneto				CW
Gun Synchronizer	845			L-CW; R-CCW
Hydraulic Pump	3040			CW
Fuel Pump	1900			CCW
Vacuum Pump	3040			CW
Generator	3040			CW
Tachometers	1900			L-CW; R-CCW
Propeller Governor	2280			CW
Starter	3800			CCW
Remote Gear Box	3040			CW

The accessory drive torque capacities will meet the requirements of Spec. AN-9500.

D-36a. Starter.- The starter mounting pad diameter shall be seven inches. The direction of rotation when looking at the starter dog attached to the engine shall be anticlockwise, and the number of jaws on the starter dog shall be three. The starter dog shall be detachable. The clearance provided for the starter shall be in accordance with the contractor's installation drawing.

D-36b. Generator.- A standard 6 inch diameter, 6 stud mounting pad with involute spline drive will be supplied with on the engine.

D-36c. Power Take-off Drive.- The generator drive will be suitable for use as a Type II gear box drive.

D-36d. Fuel Pump.- A standard square 4 stud mounting pad with involute spline drive will be furnished on the engine, in accordance with Specification AN-9519, Figure 1.

D-36e. Gun Synchronizer Impulse Generator.- Two Type I gun synchronizer impulse generator drives in accordance with Specification AN-9520 will be provided on the engine.

D-36f. Vacuum and Hydraulic Mechanism Oil Pumps.- Two large Type II, vacuum pump mounts and .600 P.D. spline drives without tongue and groove coupling will be furnished on the engine.

D-36g. Tachometer Drives.- One each of Type I and Type II mounting pads and drives will be provided on the engine. Although clearance will be provided for an electric tachometer, the contractor will not assume responsibility for the tachometer drive assembly when the electric tachometer generator is mounted directly on the threaded outlet.

F-2c(2) Cooling Liquid.- The cooling liquid outlet temperature shall be 121°C. (250°F.).

H. NOTES

H-1 The engine manufacturer reserves the right to change the rated speed, compression ratio, carburetor, supercharger impeller diameter and gear ratio, spark plugs, and other parts of the design which affect volumetric efficiency and/or combustion efficiency, in order to meet the guaranteed engine performance, provided the airplane performance is not affected. These changes will be incorporated I the type test engine and will be made a definite part of the final coordinated specification prior to release by the Materiel Division.

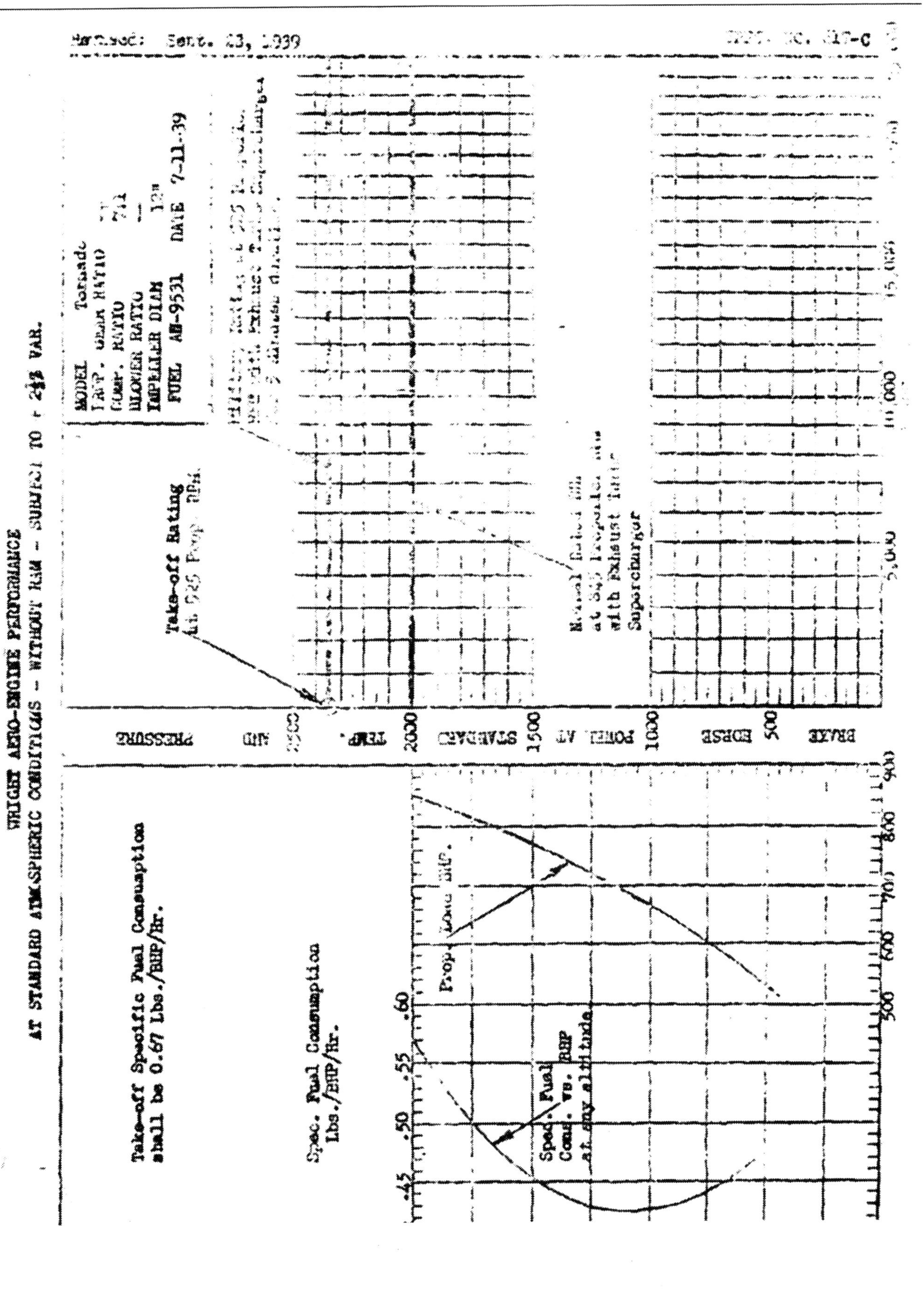
WRIGHT AERO-ENGINE PERFORMANCE
AT STANDARD ATMOSPHERIC CONDITIONS - WITHOUT RAM - SUBJECT TO + 2½% VAR.
Sept. 23, 1939
MODEL Tornado
COMP. RATIO
BLOWER RATIO
IMPELLER DIAM 12"
FUEL AN-9531 DATE 7-11-39
Take-off Rating
Normal Rated
with Exhaust
Supercharger
BRAKE HORSE POWER AT STANDARD TEMP. AND PRESSURE
500
1000
1500
2000
5,000
10,000
15,000
Take-off Specific Fuel Consumption shall be 0.67 Lbs./BHP/Hr.
Spec. Fuel Consumption Lbs./BHP/Hr.
Prop. Load BHP.
Spec. Fuel Cons. vs. BHP at any altitude
.45
.50
.55
.60
500
600
700
800
900

WRIGHT AERONAUTICAL CORPORATION

PATERSON, N. J.

U. S. A.

Date: March 21, 1941

Revised; September 11, 1941

W.A.C. ENGINE SPECIFICATION NO. 788-A

WRIGHT TORNADO ENGINE

MODEL R-2160-5 GEARED

FOR

U. S. ARMY AIR CORPS

CONSTRUCTED IN ACCORDANCE WITH

SPECIFICATION NO. AN-9500-a

DATED MARCH 30, 1940.

APPLICABLE ONLY TO AIR CORPS REQUEST FOR DATA

NO. R-40B

DATED JANUARY 29, 1940

(W.A.C. Model Designation No. 788T42AAI1)

MODEL SPECIFICATION FOR

ENGINE, AIRCRAFT

GEARED R-2160-5

FOR USE WITH EXHAUST TURBO SUPERCHARGER

A. APPLICABLE SPECIFICATIONS

A-1. The specifications shown below shall form a part of this specification except as modified herein and by the appended list, page No. 9.

AN-9500-a Engines, Aircraft, General Specification.

AN-9501-a Engines, Aircraft, Model Specification.

B. TYPE

B-1. This specification covers the requirements for the R-2160-5 engine.

C. MATERIAL AND WORKMANSHIP

C-1. The requirements of Specification AN-9500-a, Par. C-1 shall be applicable except that the words, "and contain provisions for adequate tests", shall be deleted.

D. GENERAL REQUIREMENTS

D-1. See Section E.

E. DETAIL REQUIREMENTS

E-2. Drawings. - The following Wright 'Aeronautical Corporation drawings form part of this specification:

Dwg. No. **Drawing Title**

414670 – Engine Assembly, Basic (showing accessory drive oil seals)
414675 – installation Drawing (showing clearances for engine accessories and their removal)
413461 – Priming System Assembly
413462 – Screen Assembly, Air Inlet
110559 – Gasket, Screen, Air Inlet
112260N1, 110560N1, 68973N30 – Carburetor, Stromberg, Picture, Assembly & Setting.
112850 – Spark Plug Assembly
114092N, 114093N – Terminal, Spark Plug
414675 – Installation, Cylinder Cooling Baffles, *Lubrication System Diagram, *Clearance Chart
114092N – Ignition Assembly, Scintilla Radio Shielded -Intake Side
114093N – Ignition Assembly, Scintilla Radio Shielded-Exhaust Side
*Included in Instruction Book.

E-3. Acceptance. - Acceptance of this engine as a service type shall be predicated on prior acceptance of the R-11-160-1 engine in accordance with Spec. No. XC-28398.

E-3(I). (Ref. AN-9503-a, F-6a(1).) The requirements of this paragraph shall not be applicable.

E-4. Weight. - The total dry weight of the engine complete shall not exceed the values indicated below:

Basic engine, including integral superchargers, supercharger drive mechanism, propeller reduction gears, engine lubrication system oil pumps, starter connection, including starter dog, gun synchronizing impulse generator drive (but not impulse generators), tachometer drives, fuel pump drive, generator drive, vacuum pump drives, propeller governor drive, all piping and controls between engine parts, and torque meter (lbs)	2582.00
Carburetor and Adapter	50.00
Carburetor Screens and Gasket	2.00
Magnetos & Ignition Distributors	105.00
Radio Shielded Ignition Assembly complete with cable	40.00
Spark Plugs	12.00
Priming System on Engine	2.00
Cooling Air Deflectors and Baffles	0.00
Accessory Drive Covers	2.00
TOTAL DRY WEIGHT OF ENGINE	2735.00

E-5. Performance Characteristics. - The ratings specified herein, and the curves specified herein and shown on pages 7 and 8 shall constitute the power and specific fuel consumption guarantees. The terms used and the standard conditions shall be in accordance with the applicable definitions contained in Specification AN-9502a.

E-5a. Ratings. - The engine shall be rated as follows, using fuel conforming to Specification AN-VV-F-781 and oil conforming to Specification AN-9532, Grade 1120 and an exhaust turbo supercharger of suitable capacity;

2000 BHP at 3800 RPM at Sea Level
2000 BHP at 3800 RPM at 25,000 feet, normal rated altitude.
2350 BHP at 4150 RPM, Take-off for five minutes.
2350 BHP at 4150 RPM military rating at 25,000 feet, military rated altitude.
4270 RPM Rated overspeed dive RPM.

E-5a(1). When necessary, demonstration of the guaranteed performance specified under E-5a shall be made on the dynamometer under simulated altitude conditions and with best power mixtures, except that take-off power shall be demonstrated with the take-off specific fuel consumption specified on page 7 of this specification.

E-5e. Specific Oil Consumption. - The specific oil consumption shall not exceed 0.035 Lb./BHP/Hr. at normal rated power and speed, and 0.025 Lb./BHP/Hr. at 70% normal rated power and 89% normal rated speed.

E-5l. Oil Flow and Heat Rejection, - Heat rejection data will be furnished after completion of test specified in Paragraph E-3 above.

E-6, The complete engine shall function satisfactorily up to 30,000 feet. Satisfactory operation to the service ceiling of the airplane shall be provided if possible.

E-7. Propeller. - The engine shall have a number 60 propeller shaft end. Provision shall be made for a governor typo propeller mechanism.

E-12. Overall Dimensions. - The overall dimensions of the basic engine shall not exceed the following:

Length	95.07 Inches
Diameter	36.50 Inches

E-14. Preparation for Storage. - The engine shall be prepared for storage prior to shipment, in accordance with Specification AN-9505a and Amendment No. 1 except that Air Corps Specification 3368-B material shall be used in lieu of AN-9508 material specified and the first sentence of Paragraph E-3e of Specification AN-9505a shall be deleted.

E-18. Propeller Drive. - The engine shall be equipped with a reduction gear ratio of .25. The direction of propeller rotation when viewed from the anti-propeller end shall be anti-clockwise.

E-19. Impeller Gear. - The impeller gear ratio shall be 4.20:1 and the impeller shall be 12 inches in diameter.

E-20. Pistons. - The engine shall be fitted with pistons of 7.0:1 compression ratio.

E-23a(l). Spark Plugs. - The engine will be fitted with Bendix BG-466S, 14 mm. spark plugs in accordance with contractor's drawing No. 112850.

E-23b. Radio Shielded Ignition Assemblies. - The engine shall be equipped with radio shielded ignition assemblies in accordance with contractor's drawing Nos. 114092N and 114093N.

E-23d. Magnetos. - The engine shall be equipped with Scintilla DF-6L magnetos in accordance with contractor's drawing No. 112174N1.

E-23g. Cooling, - Provisions will be made for cooling the interconnecting parts of the ignition system. Temperature requirements will be met provided adequate cooling air is supplied.

E-24c. Oil Leakage. - With a mixture of equal parts of aviation gasoline and Grade 1100 oil in accordance with Specification AN-9532, at ordinary room temperatures, supplied to the oil pump inlet under a head of six feet, the total flow of oil into the pump due to leakage of oil into the engine shall not exceed 0.3 pound per hour

E-24c. Scavenging. - The scavenging pump shall adequately scavenge the engine for extended-periods of time under all normal operating conditions provided the back pressure on the scavenge system does not exceed 30 Lbs./Sq. In. at rated power and speed, decreasing uniformly with engine speed to 10 Lbs./Sq. In. maximum at idle.

E-24g Oil Cleaner - Foreign matter being removed by the cleaner shall not reenter the lubricating system under normal operating conditions.

E-24j. Oil Connections. - Oil inlet and outlet connections will be supplied on the engine.

E-24q. Crankcase Breather. - The breather shall terminate in a hose nipple connection in accordance with manufacturer's installation drawing. The *opening will* normally be covered by a breather cap.

E-25. Fuel Metering System. - The engine shall be equipped with a Stromberg PR-55A1 carburetor in accordance with W.A.C. Drawing Nos. 110560N1 & 112260N1.

E-25(l). The requirements of Specification AN-9515-a pertaining to carburetor setting determination shall not be applicable. An experimental carburetor setting shall be provided which will deliver, in the auto rich position, mixtures within +4% -0% of the guaranteed specific fuel consumption table shown on page No. 8. Carburetor setting data will be provided to the Procuring Agency prior to flight operation of the engine.

E-26. Fuel Priming System. - The requirements of Specification AN-4033 shall not be applicable. A priming system shall be supplied by the contractor satisfactory for normal cold weather starting. A solenoid priming valve shall be supplied for mounting on the carburetor.

E-32. Exhaust System. - The engine shall not be equipped with a hot spot.

E-32a&b. Exhaust Flanges. - Exhaust flanges shall be furnished with the engine but shall not be included in the weight of the engine.

E-36. Accessory Drives. - The speed of each accessory drive at normal rated engine speed, the maximum permissible torque in inch pounds for continuous operation, the maximum permissible static torque in inch pounds and the direction of rotation when looking at the end of the accessory drive shaft in the engine shall be as follows. Accessory drives shall not overspeed at any normal rated engine speed except as modified by the drive ratios in the following list of accessory drive ratios:

Accessory Drive	Ratio	Maximum Continuous Torque (Inch Lbs)	Maximum Static Torque (Inch Lbs	Rotation
Magneto	0.750:1			CW
Gun Synchronizer	0.250:1	20	200	CW
+Hydraulic Pump	0.800:1	150	2250	CW
Fuel Pump	0.474:1	150	450	CCW
+Vacuum Pump	0.800:1	150	2250	CW*
Generator (Hi-Speed)	1.775:1	400	4000	CW
Tachometers (2)	0.500:1	4	50	CCW
Propeller Governor	0.645:1	65	600	CCW
Starter	1.000:1		36000	CCW
*Power Take-off	0.800:1	1200	12000	CW

*Applicable to 24 Tooth Spline Only.

+Applicable to 12 Tooth Spline Only.

E-36a. Starter. - The starter mounting pad and drive shall be Type I in accordance with Specification AN-9517. The direction of rotation when looking at the starter dog attached to the engine shall be anti-clockwise. Sufficient clearance shall be provided for the installation and removal of an Eclipse E-160 or Series 41 starter or other starter having equivalent dimensions.

E-36b. Generator. (Hi-Speed). - The Hi-Speed generator mounting pad and drive shall be in accordance with Fig. 1 of Specification AN-9518.

E-36c . Power Take-off Drive. - The low speed generator drive will be suitable for use as a Type II gear box drive.

E-36d. Fuel Pump. - The fuel pump will be mounted on the oil pump but will not be driven from the oil pump.

E-36e. Gun Synchronizer Impulse Generator. - Two Type I gun synchronizing impulse generator drives in accordance with Specification AN-9520 will be provided on the engine.

E-36f. Vacuum and Hydraulic Mechanism Oil Pumps. - Two large, Type II, vacuum pump mounts and drives will be furnished on the engine.

E-36g. Tachometer Drives. - One each of Type I and Type II mounting pads and drives will be provided on the engine.

F. <u>METHOD OF SAMPLING, INSPECTION, AND TESTS</u>

F-1. (Ref. AN-9503-a, Par. F-6f (6) Cooling Liquid. - The cooling liquid outlet temperature shall be 121°C. (250°F.).

H. NOTES

H-1. The engine manufacturer reserves the right to change the rated speed, compression ratio, carburetor, supercharger impeller diameter and gear ratio, spark plugs, and other parts of the design which affect volumetric efficiency and/or combustion efficiency, in order to meet the guaranteed engine performance, provided the airplane performance is not affected. These changes will be incorporated in the type test engine and will be made a definite part of the final coordinated specification prior to release by the Materiel Division.

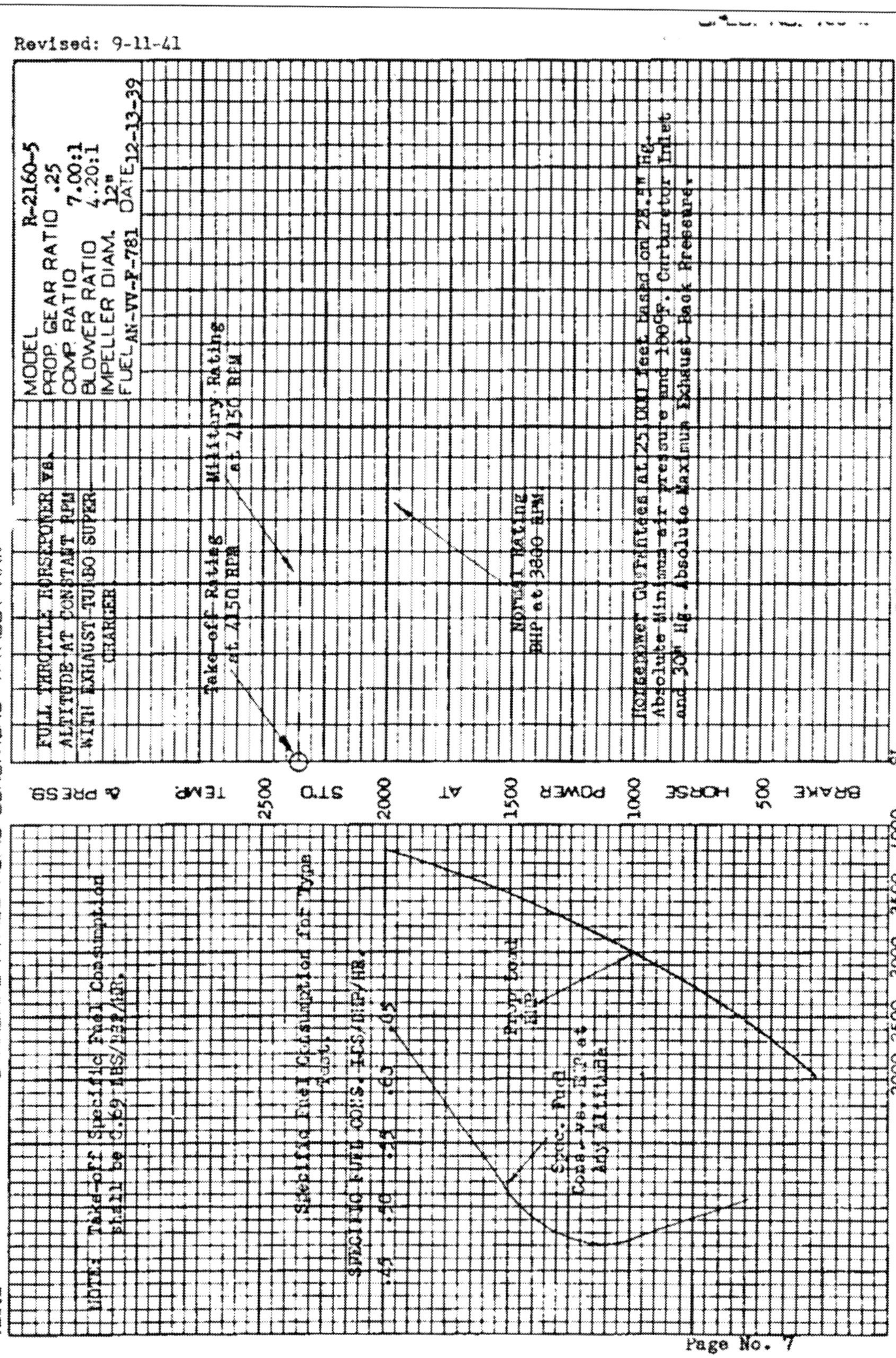
Revised: 9-11-41
WRIGHT AERO-ENGINE PERFORMANCE
AT STANDARD ATMOSPHERIC CONDITIONS - WITHOUT RAM
MODEL R-2160-5
PROP. GEAR RATIO .25
COMP. RATIO 7.00:1
BLOWER RATIO 4.20:1
IMPELLER DIAM. 12"
FUEL AN-VV-F-781 DATE 12-13-39
FULL THROTTLE HORSEPOWER vs. ALTITUDE AT CONSTANT RPM WITH EXHAUST-TURBO SUPER-CHARGER
Take-off Rating at 4150 RPM
Military Rating at 4150 RPM
Normal Rating BHP at 3800 RPM
Horsepower Guarantees at 25,000 Feet based on 28.5" Hg. Absolute-Minimum air pressure and 100°F. Carburetor Inlet and 30" Hg. Absolute Maximum Exhaust Back Pressure.
NOTE: Take-off Specific Fuel Consumption shall be 0.69 LBS/BHP/HR.
Specific Fuel Consumption for Type Test.
SPECIFIC FUEL CONS. LBS/BHP/HR.
.45 .50 .55 .60 .65
Prop. Load HP
Spec. Fuel Cons. vs. BHP at Any Altitude
BRAKE HORSE POWER AT STD TEMP & PRESS.
500 1000 1500 2000 2500
2000 2500 3000 3500 4000
CRANKSHAFT RPM
S.L. 5,000 10,000 15,000 20,000 25,000
ALTITUDE - FEET
Page No. 7

-124

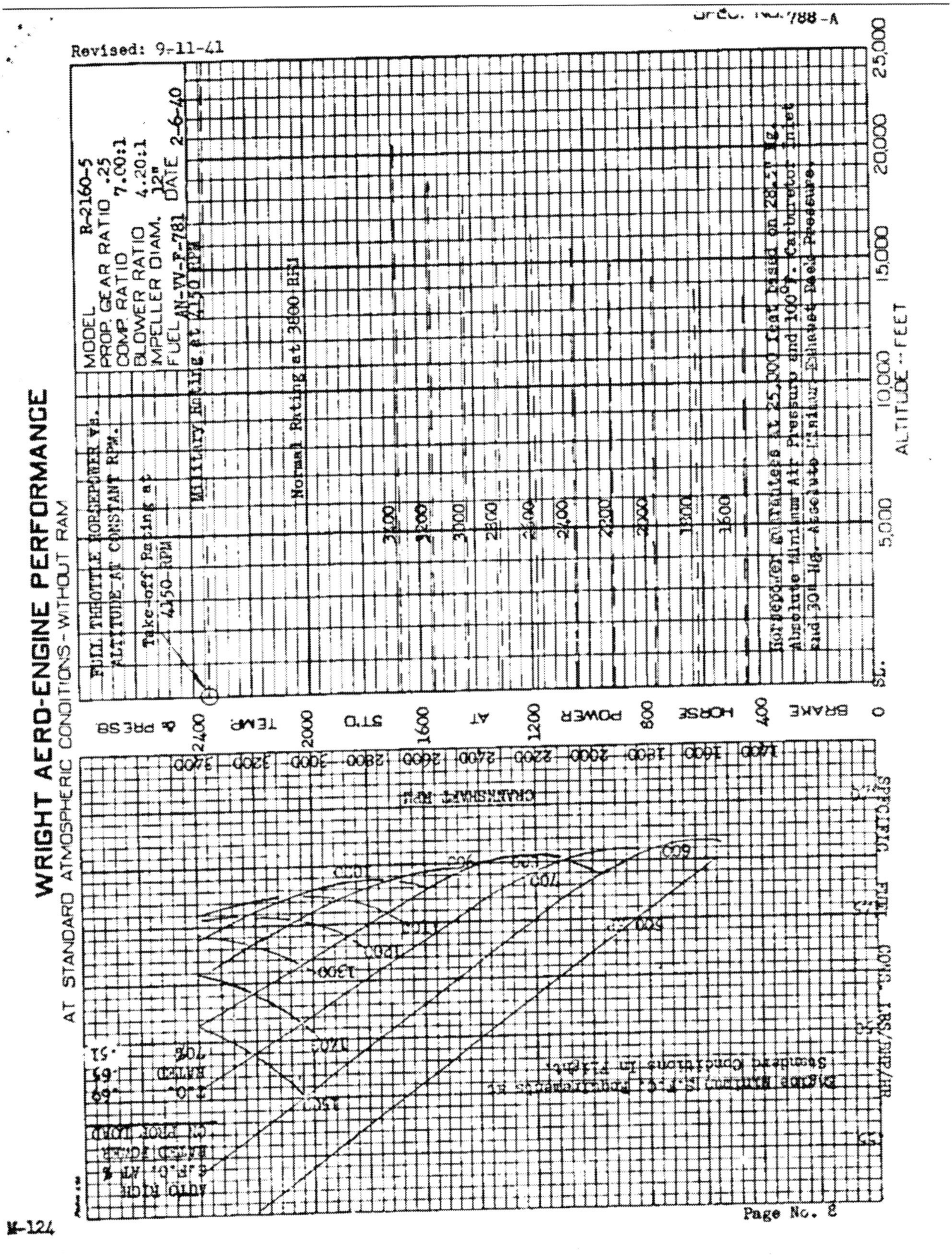
WRIGHT AERO-ENGINE PERFORMANCE
AT STANDARD ATMOSPHERIC CONDITIONS - WITHOUT RAM
Revised: 9-11-41
SPEC. NO. 788-A
FULL THROTTLE HORSEPOWER vs. ALTITUDE AT CONSTANT RPM.
MODEL R-2160-5
PROP. GEAR RATIO .25
COMP. RATIO 7.00:1
BLOWER RATIO 4.20:1
IMPELLER DIAM. 12"
FUEL AN-VV-F-781
DATE 2-6-40
Take-off Rating at 4150 RPM
Military Rating at 4150 RPM
Normal Rating at 3800 BHP
Horsepower guaranteed at 25,000 feet based on 28.5" Hg. Absolute Minimum Air Pressure and 100°F. Carburetor Inlet and 30" Hg. Absolute Minimum Exhaust Back Pressure.
ALTITUDE--FEET
BRAKE HORSE POWER AT STD. TEMP. & PRESS.
CRANKSHAFT RPM
Standard Conditions in Flight
Page No. 8
M-124

Appendix E—Data from Engine No. 11770

R-2160-3 Data taken from Engine No. 11770 at the New Jersey Aviation Hall of Fame and Museum, Teterboro, New Jersey

Engine Data Plate

Air Corps or Bureau Model	R-2160-3
Air Corps or Bureau Number	44-4393
Manufacturers	11770
Contract Number	W535AC13709
Engine Spec Number	689
Propeller Rotation	Right
Prop Gear Ration	4.33:1
Compression Ratio	7.0:1
Cylinder Number	A-123456
Impeller Diameter	12"
Impeller Ratio	4.2:1
Oil Pressure	45 PSI

Timing Data:	
Spark Advance, Degrees	
Intake	30 BTDC
Exhaust	30 BTDC
Valve Timing, Degrees	
IO	41 BTC
IC	63 ABC
EO	74 BBC
EC	66 ATC
Valve Gear Timing, Engine Cold	
Intake	0.020
Exhaust	0.020
Reset	
Intake	0.010
Exhaust	0.015

Firing Order:

A1-C4-D5-F2-G3-B6
C1-E4-F5-A2-B3-D6
E1-G4-A5-C2-D3-F6
G1-B4-C5-E2-F3-A6
B1-D4-E5-G2-A3-C6
D1-F4-G5-B2-C3-E6
F1-A4-B5-D2-E3-G6

Nose Section

Electric Propeller Alternator	
Model Number	124512
Serial Number	9281
Ignition Booster Coil	
Point adjustment	1.3 Amps at 13 Volts
Part Number	10-24425-1
Serial Number	0003
Magnetos, Bendix Scintilla DF6L	
Specification	112174
Part Number	10-21100-1
Date of Manufacture	4-4-40
Harness, Exhaust, Scintilla Assembly Radio Ignition	
Type	T-42
Part Number	10-27850
Harness, Intake, Scintilla Assembly Radio Ignition	
Type	T-42
Part Number	10-27625

Magneto and Ignition Harness Serial Numbers

"A" Bank	
Magneto	0410
Intake Harness	12297
Exhaust Harness	12278
"B" Bank	
Magneto	0416
(this Model Plate has the "6" in DF6L stamped upside down to read DF9L)	
Intake Harness	12302
Exhaust Harness	12276
"C" Bank	
Magneto	0414
Intake Harness	12171
Exhaust Harness	12282
"D" Bank	
Magneto	0415
Intake Harness	12309
Exhaust Harness	can not read because of oil sump
"E" Bank	
Magneto	0413
Intake Harness	can not read because of oil sump
Exhaust Harness	12279
"F" Bank	
Magneto	0412
Intake Harness	12306
Exhaust Harness	12277
"G" Bank	
Magneto	data plate unreadable
Intake Harness	12209
Exhaust Harness	data plate unreadable

Accessory Case

Injection Carburetor, Bendix Aviation Corp, South Bend, Indiana

Model	PR-55A1
Serial number	04141
Parts List	392098-4
Assy Dwg No	<is blank>
Eng Mfg Set Number	68973N-730
Fuel Pressure	15 psi

Aviation Engine Air Pump, Eclipse Aviation

Model	550
Model	40xxx
Style	A
Maximum Pump rpm	4000
Serial Number	B-417
Mfg Dwg	D-82306

Tachometer Generator, Small Motors Inc., Chicago, Ill

MS-25038-1(A5G)	
MIL-G-54138(A5G)	
Stock No.	6680-713-4522
Serial Number	AF-59-3661-12
Part Number	GTC3-025A
Contract No	AF33(600) –39659.

Hydraulic Pump, Vickers

Model Number	PF17L-3911-25Z
Serial Nubmer	MX2097P
PSI	3000
Assy No.	136143
RPM	3000
Cu In Per Revolution	506

Aircraft DC Generator

Type No.	30E07
Model No	7
Mfg No.	30E07-7-B
Serial Number	104
Style	D
RPM	3000-8000
Volts	30
Amps	200

Starter, Aircraft, Direct Cranking For 28VDC, Jack & Heintz

AN11168-3	
Clutch Setting	850 lb ft
Rotation	Right
Mfg S/N	unreadable
Mfg P/N	JH11900R3
Mfg M/N	JN-6FR3

No data plate on left generator.

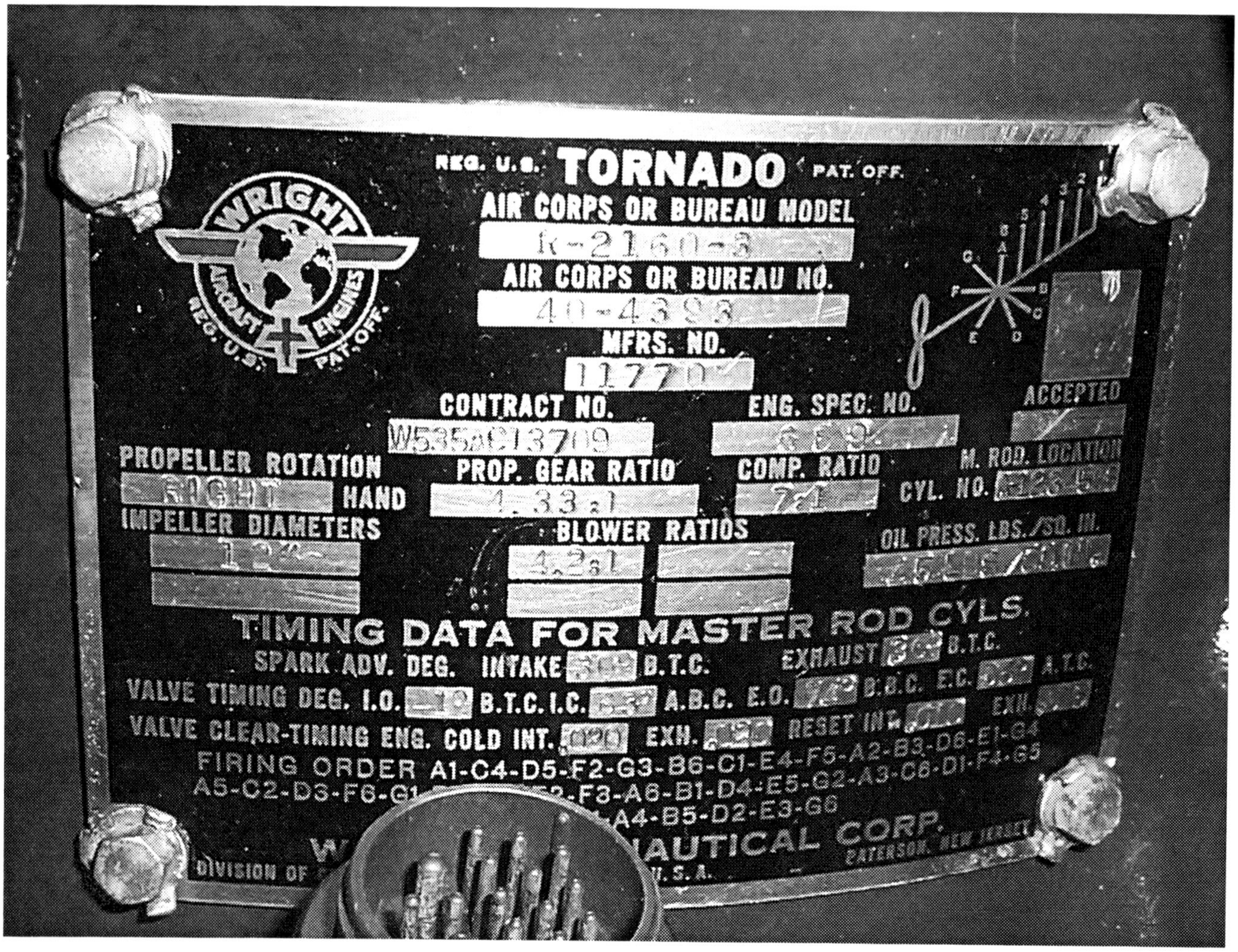

The Tornado 42 Data Plate

Appendix F—Hughes D-2 Installation Engineering Details

CONFIDENTIAL

COMPARABLE FEATURES ON T42 ENGINE INSTALLATIONS 4/16/41

Items	Hughes D-2	Lockheed XP-58
Engine Nacelle	50" wide, 60" deep	54" x 66" - subject to slight revision.
Propeller	4 blade Ham.Std. 15'2", 24E60 Hub, 624 BA-O Blade	4 interchangeable choices.15'2" 1. Ham.Std. 6483 Blade 24D60 Hub. 2. Ham.Std. 6243 Blade 24E60 Hub. 3. Curtiss 924 Blade. 4. " 1018 Blade.
Spinner	41" dia. about 36" long.	Furnished by prop manufacturer to contour supplied.
Oil Cooler	2 - 15" dia., 9" core.	2 elliptical, 175 sq.in. each, 9" core.
Prestone Cooler	2 - 259' each, 9" core in L.E. wing.	Single 5½ sq.ft. elliptical 9" core.
Turbos	2 G.E. in parallel Type B-2 1700/25000	1 G.E. type - XE 2500 HP/ 25,000'
Turbo Oil Supply	Separate system	Separate system
Turbo Regulator Oil Supply	Desire engine oil	Turbo oil system
Coolant Lines	3 inch diameter	Tentatively 2" diameter.
Installation engine Delivery Date	?	October 15, 1941
Actual Engine Delivery	?	April 15, 1941
Preheat	100° Rise	100° Rise + alcohol

John McElroy

JM:ik

CONFIDENTIAL

ACCESSORIES

I. ATTACHED TO OR PART OF ENGINE

	Type	Model	Manufacturer	Drive No.*
Carburetor		PR-52 ?	Stromberg	
Magneto	Dual	DF-6L	Scintilla	
Booster				
Starter	602	M-15	Eclipse	
Spark Plugs	14 mm	BG-466S	Bendix	
Shielding				
Generator	AC-02	½ Speed	Eclipse	16 Tooth
Oil Filter				
Hydro Valve	—	—		
Prop. Governor Pump	Woodward	SH8-01A	Ham. Std.	
Gun Synchronizer	None			
Tachometer				
Fuel Pump	CWB	R-600	Pesco	
Vacuum Pump	B-8	211-J	Pesco	12 Tooth
Hydraulic Pump			Vickers	16 Tooth
Supercharger Regulator	N-1		G.E.	
Primer Solenoid			Bendix	
Gurgle Tube (Turbo?)	Turbo		WAC	
Torquemeter			WAC	
Alternator				
Cabin Supercharger	Roots			Std Gen. Dr.

Accessory Drive Type ______ SU-Dwg. No. ______ Ratio ______

* Note: List drive number and type of drive: Square Spline, Involute Spline, or tongue.

II. RELATED TO OPERATION BUT NOT ATTACHED TO ENGINE

A. FUEL SYSTEM

	Type	Model	Manufacturer
Relief Valve			
Bypass Valve			
Flow Meter	RU-100		Eclipse
Wobble Pump			
Primer Pump			
Strainer			
Dump Valve			
Combination Unit			
Primer Nozzle			
Remote Fuel Pump	24 U	X 308-C	Pesco
Primer Pump Location			
Hydraulic Generator	Fuel Pump	248	Pesco

OIL SYSTEM

	Type	Model	Manufacturer
Oil Cooler	2-14" Parallel Flow		UAP
Oil Cooler Relief Valve			
Oil Temperature Regulator	F-5	U-3820	UAP
Oil Cooler Shutters		Before or After Cooler	
Oil Cleaner (Cuno)	Engine Strainer		
Oil Shutter Control	Location		
Dilution Valve			
Fire Shut Off Valves			
Drain Valve	Y 1½"	U-1060-3	UAP

	D-2	XP-58
Nacelle	50" wide, 60" deep	
Prop	4 blade Hydromatic 15'2" 24E60 Hub, 6274A-0 Blade	
Oil Cooler	2-15" 9" core	2 elliptical 175 sq" - 9" core
Prestone Cooler	2 - 4 square feet. Total 9" core	5½ sq.' elliptical 9" core
Turbos	2 G.E. B-2	1 G.E. Model XE 2500/25000'
Spinner	Hughes 41" dia about 36" long	Prop manufacturer
Intercoolers		Harrison, Airesearch, or Youngs
Coolant Line		2" 12#/sq" back press. 70% water.
Installation Engine Actual Engine		On or before Oct 15 Apr. 15.
Turbo Oil Supply	Separate system - Engine oil for regulator 1.5 pints/min.	Separate system - Turbo oil for regulator.
Carb. Heat	100° rise	40° rise + alcohol

OIL SYSTEM

	Type	Model	Manufacturer
Oil Cooler	2-14" Parallel Flow		UAP
Oil Cooler Relief Valve			
Oil Temperature Regulator	F-5	U-3820	UAP
Oil Cooler Shutters		Before or After Cooler	
Oil Cleaner (Cuno)	Engine Strainer		
Oil Shutter Control	Location		
Dilution Valve			
Fire Shut Off Valves			
Drain Valve	Y 1½"	U-1060-3	UAP

[illegible] straight".

	D-2	XP-58
Nacelle	50" wide, 60" deep	[illegible]
Prop	4 blade Hydromatic 15'2" 24E60 Hub, 6274 BA-0 Blade	[illegible]
Oil Cooler	2-15" 9" core	2 elliptical 175 sq" - 9" core
Prestone Cooler	2 - 4 square feet Total, 9" core	5½ sq.' elliptical 9" core
Turbos	2 G.E. B-2	1 G.E. Model XE 2500/25000'
Spinner	Hughes 41" dia about 36" long	Prop manufacturer
Intercoolers		Harrison, Airesearch, or Youngs
Coolant Line		2" 12 #/sq" back press. 70% water.
Installation Engine Actual Engine		On or before Oct 15 Apr. 15.
Turbo Oil Supply	Separate system Engine oil for regulator 1.5 pints/min.	Separate system - Turbo oil for regulator.
Carb. Heat	100° rise	40° rise + alcohol

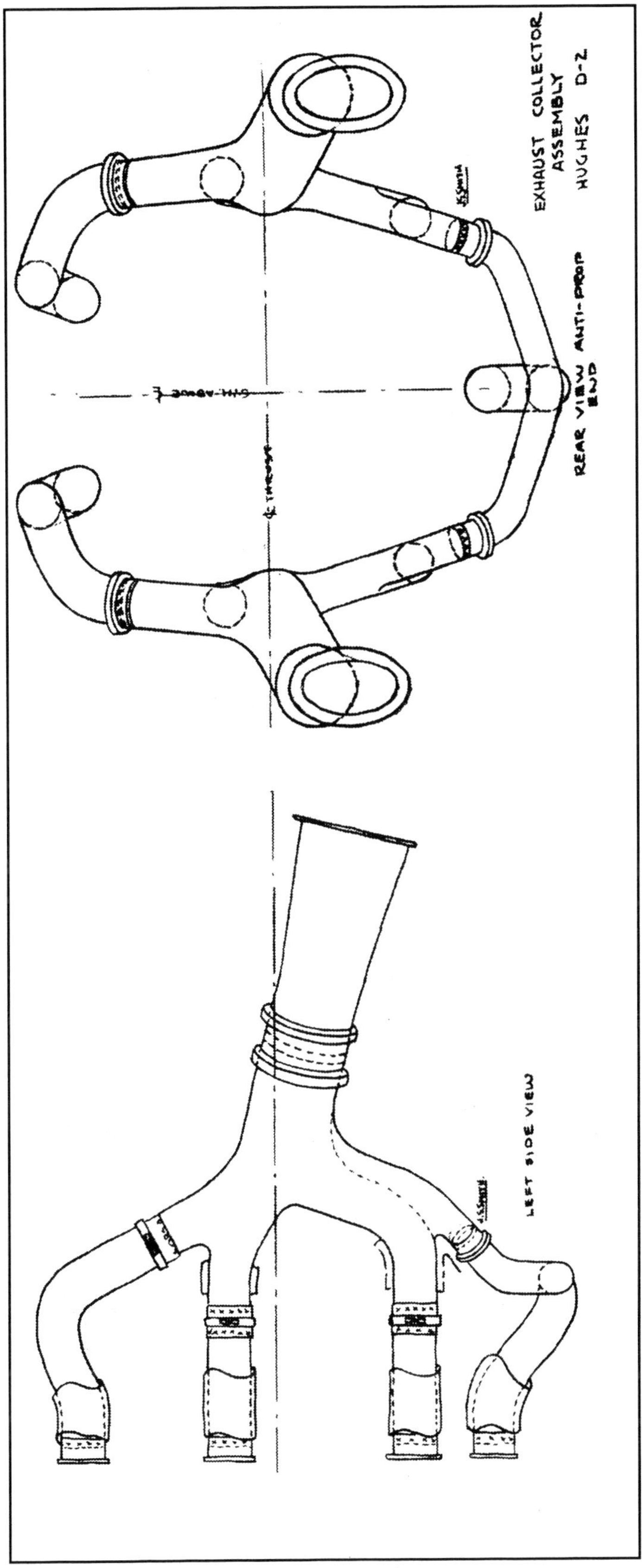
EXHAUST COLLECTOR
ASSEMBLY
HUGHES D-2
REAR VIEW ANTI-PROP
END
LEFT SIDE VIEW

Appendix G—Lockheed XP-58 Installation Engineering Details

CONFIDENTIAL

POWER PLANT

ITEM	WEIGHT
Nacelles	847.0
Engine Mounts	430.0
Cowling	336.0
Firewalls	38.0
Protective Coating	1.0
Misc.	7.0
Rivets, Nuts & Bolts	19.0
Prov. for - Oil	14.0
- Anti-Icing	2.0
Engines	5,470.0
Supercharging System	1,750.0
Turbos	800.0
Lube System	57.0
Supports & Misc.	48.0
Controls	47.0
Exhaust Sys. - F.W. Aft	
Pipe	146.0
Shroud	95.0
Waste Jet	19.0
Carb. Hot-Air Duct	24.0
Exp. Joints	12.0
Brack., Supt. & Misc.	84.0
Intercoolers & Supts.	250.0
Air Ducts	95.0
Oil Coolers	73.0
Exhaust System	402.0
Exhaust Stacks	59.0
Collecting Manifold	69.0
Shroud-Coll. Man.	32.0
Pipe-Coll. Man to Ring	15.0
Shroud-Coll. Man. to Ring	11.0
Collector Ring	38.0
Shroud - Coll. Ring	54.0
Pipe - Ring to Firewall	57.0
Shroud - Ring to F.W.	36.0
Exp. Joints	4.0
Brack., Supt. & Misc.	27.0

ITEM	WEIGHT
Air Intake System	189.0
Intake to Turbo	18.0
Turbo to Intercooler	18.0
Intercooler to F.W.	76.0
Fwd. of F.W.	7.0
Joints	16.0
Brack, Supts & Misc.	54.0
Vacuum Pump & Separator	6.0
Propellers	1,203.0
Blades)	
Hubs)	1,060.0
Spinners	79.0
Governors	11.0
Pitch Control	17.0
Feathering Pumps	12.0
Pump Lines	5.0
Feath. Oil Lines	5.0
Shut-Off Valves	1.0
Residual Oil	3.0
Misc.	10.0
Prop. Slinger Rings	15.0
Starting System - Primer	2.0
Cooling System	1,593.0
Radiators	427.0
Shutters	28.0
Scoops & Ducts	133.0
Exp. Tank	19.0
Liq. in - Engine	260.0
- Radiator	160.0
- Exp. Tank	40.0
- Piping	380.0
Piping	100.0
Supports & Instal.	38.0
Misc.	8.0

CONFIDENTIAL

6

CONFIDENTIAL

POWER PLANT (Cont'd.)

ITEM	WEIGHT
Lubrication System	973.0
Oil Coolers - Incl. Supt.	403.0
Supports	10.0
Ducts	180.0
Oil Tanks - Incl. Supports	80.0
Piping	19.0
Misc.	17.0
Residual Oil	264.0
Intermediate Booms	449.0
Bulkheads & Frames	26.0
Tubular Structure	234.0
Fittings	52.0
Gussets	4.0
Covering	88.0
Protective Coating	4.0
Misc.	10.0
Rivets, Nuts Bolts	20.0
Prov. for - Super.	8.0
- Surf. Controls	3.0
Radiator Shutter Controls	24.0
Oil Cooler Shutter Controls	28.0
Superchgr. Oil Cooler Controls	12.0

CONFIDENTIAL

CONFIDENTIAL

COMPARABLE FEATURES ON T42 ENGINE INSTALLATIONS 4/16/41

Items	Hughes D-2	Lockheed XP-58
Engine Nacelle	50" wide, 60" deep	54" x 66" - subject to slight revision.
Propeller	4 blade Ham.Std. 15'2", 24E60 Hub, 624 BA-0 Blade	4 interchangeable choices. 15'2" 1. Ham.Std. 6483 Blade 24D60 Hub. 2. Ham.Std. 6243 Blade 24E60 Hub. 3. Curtiss 924 Blade. 4. " 1018 Blade.
Spinner	41" dia. about 36" long.	Furnished by prop manufacturer to contour supplied.
Oil Cooler	2 - 15" dia., 9" core.	2 elliptical, 175 sq.in. each, 9" core.
Prestone Cooler	2 - 259' each, 9" core in L.E. wing.	Single 5½ sq.ft. elliptical 9" core.
Turbos	2 G.E. in parallel Type B-2 1700/25000	1 G.E. type - XE 2500 HP/ 25,000'
Turbo Oil Supply	Separate system	Separate system
Turbo Regulator Oil Supply	Desire engine oil	Turbo oil system
Coolant Lines	3 inch diameter	Tentatively 2" diameter.
Installation engine Delivery Date	?	October 15, 1941
Actual Engine Delivery	?	April 15, 1941
Preheat	100° Rise	100° Rise + alcohol

John McElroy

JM:ik

CONFIDENTIAL

CONFIDENTIAL

At Los Angeles, Calif.

Report No. 3

May 25, 1942

Reg #7873

CC Messrs: Nutt to Pierce
Sulzman to Johnson to Cannon to FEF
Moerman to Field Engineers to FEF
Brown " " " " "
Stanley " " " " "
Leak to Lundquist
Cincinnati Division
Goodman to WAL
McElroy
Central Files

PROJECT NO. F-311

Lockheed Model XP-58 - Tornado 42 Engine Intercepter Pursuit

SUBJECT: Cooling System Description.

OBJECT:

To describe the cooling system of the subject installation.

CONCLUSIONS:

We believe the cooling system described herein to be satisfactory insofar as it is possible to predict at this time.

DISCUSSION:

The XP-58 cooling system is designed along conventional Army requirements. It is described on the Lockheed drawing #194648, attached to the original of this report.

Prestone Radiator

The coolant radiator is located in the lower part of the boom after the wing. Cooling air is obtained from a scoop underneath the boom directly in front of the radiator and is discharged directly aft of the radiator at the bottom of the boom. Four exit flaps control the flow of air through the radiator. These flaps are automatically controlled by a regulator of Lockheed design which controls the hydraulic flap actuating motor.

The coolant radiator itself is a Harrison fin and tube radiator, part #A3945. It has 650 sq. in. of frontal area and a 12 in. core. The design basis of this cooler is two-engine, military power climb, at which it is estimated that 38,500 BTU can be rejected at a 250°F coolant temperature and a 388 gal. per minute coolant flow. This cooler would therefore be adequate for the 2350 HP Tornado engine, provided that not more than 38.5% of the brake horsepower is rejected to the coolant. Air flow required to reject this amount of heat is estimated to be 1800 lbs per min. at 100°F. This requires a pressure drop

Project No. F-311 CONFIDENTIAL 5/25/42
Report No. 3 Page 2

DISCUSSION: (Continued)

Prestone Radiator (Cont'd.)

of 13-1/2 in. of water across the cooler.

Coolant Lines and Connections

A 3 in. diameter coolant line from the bottom of the cooler is divided into two 2 in. diameter lines which run forward through the firewall to the engine coolant pumps. The two 1-3/4 in. coolant return lines are run from the two lower outlets on the front of the engine to the top of the cooler. Lockheed are fabricating their own 1-3/4 in. coolant outlet fittings to be installed on the engine. This is to eliminate the necessity for reducing fittings with the standard engine connections. All lines are dural tubing joined by hose connections with single hose clamps.

Thompson quick disconnect fittings are used on all coolant lines passing through the firewall. These fittings can be broken by turning a locking nut 90°. This action also seals each side of the fitting so that the coolant lines will not drain upon separation.

When the airplane is set on the ground at a 5° angle, there is a slight liquid trap in each coolant outlet line near the rear of the engine and another in the 3 in. diameter coolant inlet line just forward of the radiator. Drain cocks are installed in each of these three locations to permit complete drainage of the system. With the exception of the above traps the coolant system may be completely drained at the drain cock at the bottom of the radiator.

Expansion Tank

A positive head is maintained on each coolant pump inlet from a 4.72 gal. header tank located above the engine and connected to each coolant inlet line by a 1 in. diameter line. Four and one-half gallons of this header tank are for expansion space. Since the coolant system is filled through this tank, a standpipe arrangement at the filler cap prevents over-filling. A 3/4 in. O.D. drain line overboard to the bottom of the nacelle takes care of any overflow. The top of the radiator and the two upper vent fittings at the front of the engine are vented to this tank through a 1/2 in. O.D. tube.

Coolant system venting for change in altitude is accomplished through a 2-942-23 Parker sniffle valve at the top of the header tank. This valve is set to vent at coolant system pressures of more than 10 psi positive pressure or 3 psi negative pressure.

Coolant System Pressure Drops

The XP-58 coolant system is designed to derive the maximum benefit from the high pressures available at the coolant outlets on the engine, that is, 30 lbs per sq. in. at rated HPM. The coolant lines were reduced in diameter to raise

CONFIDENTIAL

Project No. F-311 **CONFIDENTIAL** 5/25/42
Report No. 3 Page 3

DISCUSSION: (Continued)

Coolant System Pressure Drops (Cont'd.)

the total pressure drop between the engine coolant outlet and the pump inlet to 30 lbs per sq. in. at rated power. By this means Lockheed were able to save approximately 145 lbs per airplane by reducing the amount of coolant necessary to fill the system. An Air Corps deviation was obtained to permit the use of a higher coolant drop than the 10 lbs per sq. in. allowed in the Army Handbook.

Production Airplane Cooling Provisions

It is expected that the above system will be modified somewhat for the second or so-called production airplane, pending the outcome of WAC's high pressure cooling tests. Lockheed would like to design their new cooling system to a 300°F coolant outlet temperature limit with a 60% - 40% Prestone - water solution. This would require a minimum pressure of 40 lbs per sq. in. in the system to prevent boiling. With such a system it would be possible to reduce both the size of the radiator and the amount of coolant required still further with the resulting saving in airplane weight and cooling drag.

John McElroy

JMcE:IK

Attached: 1 Lockheed Drawing #194648

CONFIDENTIAL

C O N F I D E N T I A L

At Los Angeles, Calif.

September 24, 1942

Report No. 4

Messrs.Nutt to Young to FEF
Sulzman to Johnson to FEF
R.F.Brown to FEF
UAL Engrg.
McElroy
Leak to Pierce
Lundquist
Central Files (2)

PROJECT NO. F-311

Lockheed Model XP-58 - Tornado 42 Engine Interceptor Pursuit

SUBJECT: Engine Mount Photographs.

OBJECT:

To transmit and discuss photographs of the engine mount for the subject installation.

DISCUSSION:

Photograph Y-7086 - This photograph pictures the first actual Lockheed engine mount from the right hand side, supporting the shell engine. As shown on Lockheed drawing #62760, this engine mount is 51 inches long, 43-1/8 inches wide at the firewall between centers on the lower truss, and 41-1/2 inches wide at the base of the upper truss. The maximum compression or tension load is a maximum of plus or minus 21050 lbs per engine pad. This load has an eccentricity of .040 inches. The engine mount weighs 163 lbs, including all attaching bolts, etc. This is the lightest mount per moment of overhang, that Lockheed have ever built.

Photograph Y-7087 - This view shows a close-up 3/4 view of the right hand side. The diagonal side member is pin-jointed at each end and must be removed to remove the engine from the mount. Its forward end is tapered to obtain clearance at the rear cover parting flange. The lower truss is machined from two solid billets of dural. The rest of the engine mount, of course, is steel tubing.

Photograph Y-7091 - This photograph taken directly behind the upper left hand engine mount pad shows the clearances between the engine mount and the rear cover parting flange of the engine. Clearance between this flange and the diagonal side member is 1/4 inch. The engine breather is shown sandwiched in between the side member and the top truss. The pipe plug shown just to the right of the top truss dog leg is inserted in the combustion heater connection.

Photograph Y-7093 - The right hand lower engine mount pad is shown in this photograph illustrating the close clearances at this point. Note that four attaching bolts are used on the lower mount pads, whereas six are used on the upper pads.

C O N F I D E N T I A L

Project No. F-311 **C O N F I D E N T I A L** 9/24/42
Report No. 4 Page 2

DISCUSSION: (Cont'd)

Photograph Y-7098 - This photograph pictures the top engine mount truss from above and illustrates the close clearances between the carburetor and the engine mount members. The combustion heater discharge fitting is shown in the supercharger section directly to the left of the upper diagonal member.

Photograph Y-7095 - Just enough clearance was provided between the lower truss and the engine oil pump to permit the removal of the oil strainer. This strainer can be removed from either above or below this truss if it is taken out in two sections. The lugs shown on the bottom of this truss are for supporting the two elliptical oil coolers.

These photographs illustrate about the cleanest design of engine mount that can be designed for this engine and represent considerable coordination between the Wright Company and the Lockheed Company. This type of mount with negligible offset at the engine mount pads is possible only by using a very wide base at the firewall, as Lockheed have done.

Since these photographs were taken, approximately 60% of the cowling and other parts of the installation have been installed on the shell engine.

John McElroy

JMcE:IK

Attached: Photographs: Y-7086
Y-7093
Y-7087
Y-7098
Y-7091
Y-7095
(1 Set Retained in WAL Files)

C O N F I D E N T I A L

CONFIDENTIAL

At Los Angeles, Calif.

December 7, 1942

Report No. 5

Messrs: Nutt to Young to FEF
Solsman to Johnson to FEF
R. F. Brown to FEF
CAL Engineering
McElroy
Leak to Pierce
Lundquist
Central Files (2)

DEC 1 1942
WRIGHT AERO LTD.

PROJECT NO. P-311

Lockheed Model XP-58, T-42 Engines

OBJECT:

To submit carburetor air filter information requested in Field Engineering Memorandum No. 166.

CONCLUSIONS:

1. A ramming filter scoop is not used but a plus .8 q pressure is expected ahead of the filter during filter on operation.

2. If the filter becomes clogged and blocked off, full preheat must be used, although limited intercooling is available to reduce CAT.

RECOMMENDATIONS:

Note "Conclusion #2" when reviewing drawings which will be submitted as soon as available.

DISCUSSION:

The requested information follows in the form of Field Engineering Memorandum No. 166.

1. Airplane Model - XP58, Engine Model - R2160.

2. The airplane has a filter installed in the main air intake.

3. The airplane was originally designed with a filter.

4. Only one filter design was made.

5. The filter element is a viscous impingement type. The filter make is not established but it will be either Air Maze R81, American UM180, or Farr.

CONFIDENTIAL

Project No. F-311 **CONFIDENTIAL** 12/7/42
Report No. 5 Page 2

DISCUSSION: (Cont'd)

6. The filter element area is 3.5 sq. ft.

7. The filter element is interposed in the air stream at a 70° angle of incidence.

8. The filter installation is such that the air enters the main scoop and must make a 90° turn upwards to enter the duct leading to the turbo. As the air turns upward, sand, etc., will continue its original horizontal flow and pass through a prestone radiator back out into the slipstream.

9. The filter element (made in 2 sections) is sealed with sponge rubber on the fore and aft edges and sealed with a fiber material on the sides. The hot air door has metal to metal sealing. Joints in the induction system ducting use standard pressure (15 lbs) seals.

10. In this turbo installation the filter is located in the main air intake ahead of the hot air door and turbo compressor.

11. The self-cleaning provision mentioned in "Item 8" can be closed only as engine cooling requirements dictate, since sand being ejected is mixed with the coolant radiator cooling air. Flaps behind the coolant radiator regulate cooling air flow.

12. Shrouds around the long exhaust tailpipes going to the turbo are used for preheat. Preheat available is 90°F above 40° at 65% rated power.

13. As in "Conclusion #2", if the filter becomes clogged and blocked off, full preheat must be used although limited intercooling will be available. The hot air door control and intercooler flap control are linked together.

14. The filter element is in the main air intake system. The filter element will be carried in the airplane baggage compartment and installed in the induction system as operations require.

15. It is expected that no sand will accumulate in the main air intake system.

16. No alternate air intake. One control for hot air door and intercooler flaps.

17. The filter element is not subjected to local heat from engine, exhaust, or turbo.

18. Filter removal time is 5 minutes. No results concerning pressure drop across the filter or affect on critical altitude are available since the airplane has not been flown.

R. J. Wright

FJW:IK

CONFIDENTIAL

Reg. # 27772.

CONFIDENTIAL

At Los Angeles, Calif.

April 22, 1943

Report No. 6

Messrs. Nutt-Young-FEF
Sulzman-Johnson-FEF
Brown-FEF
CAL Engineering
McElroy
Leak-Pierce
Lundquist
Central Files

PROJECT NO. F-311

Lockheed Model XP-58 Interceptor Pursuit - Tornado 42 Engines

SUBJECT: Airplane Status

OBJECT:

To review the status of the subject airplane.

CONCLUSIONS:

None.

RECOMMENDATIONS:

Bear in mind that nearly all of the T-42 power plant parts were actually built and could be used with this engine, or a revised version of this engine, for future flight testing.

DISCUSSION:

Lockheed have already received two Allison 3420 24-cylinder engines for installing in the first airplane. With these engines, the nacelle width will be increased about 10 inches. This increased width occurs just ahead of the wing leading edge and can be faired in very nicely. The installation will be somewhat similar to the Stinson 105 in appearance.

Power Rating

The 3420 take-off rating at present is 2600 BHP. Based on current successful flight operation of the 1710 in the P-38's at 1600 BHP, it is expected that take-off ratings of well over 3000 HP will shortly be obtained.

CONFIDENTIAL

Project No. F-311 C O N F I D E N T I A L 4/22/43
Report No. 6 Page 2

DISCUSSION: (Continued)

Heat Rejection

The present prestone cooler is expected to dissipate 40,000 BTU heat rejection, which will be adequate for the Allison engine. The 10,000 BTU capacity oil coolers are being changed to 12,000 BTU to take care of the oil heat rejection. This change would also have been necessary with the current T-42 oil heat rejection.

Weight

The specification weight of the 3420 is 2655 lb, but Lockheed have weighed the first engine at 2607 lb. This compares with the T-42 specification weight of 2735 lb. In spite of the lighter engine weight, the installed weight of the Allison will be approximately 500 lb more per engine. This is principally due to:

1. A heavier engine mount. While the T-42 mount has been described as the lightest mount that Lockheed has ever built, the Allison mount is looked upon as the heaviest.

2. More coolant system capacity necessary due to the low capacity engine coolant pump.

3. A heavier propeller.

4. Heavy P-38 exhaust collectors.

This increase in engine weight could undoubtedly be considerably reduced if more engineering time were available.

Status

The first airplane is approximately 80% complete and is expected to fly this September. Indicative of Army interest in this project is the fact that they have recently directed Lockheed to begin work on a second airplane.

Lockheed have stored the T-42 power plant installation parts pending disposition instructions from the Army. The T-42 shell engine has recently been shipped back to WAC.

The turbo installation sections just aft of the wing in the tail boom are nearly complete. This installation can be removed intact in approximately 1/2 hour.

The Lockheed power plant group appear to be quite pleased with the Allison engine. Their main criticisms have been in regard to the accessory section. They also would like to have an increased capacity coolant pump so that coolant line sizes could be reduced. The highest point on the engine is a 1-inch breather connection located on the reduction gear housing.

McE oh John McElroy

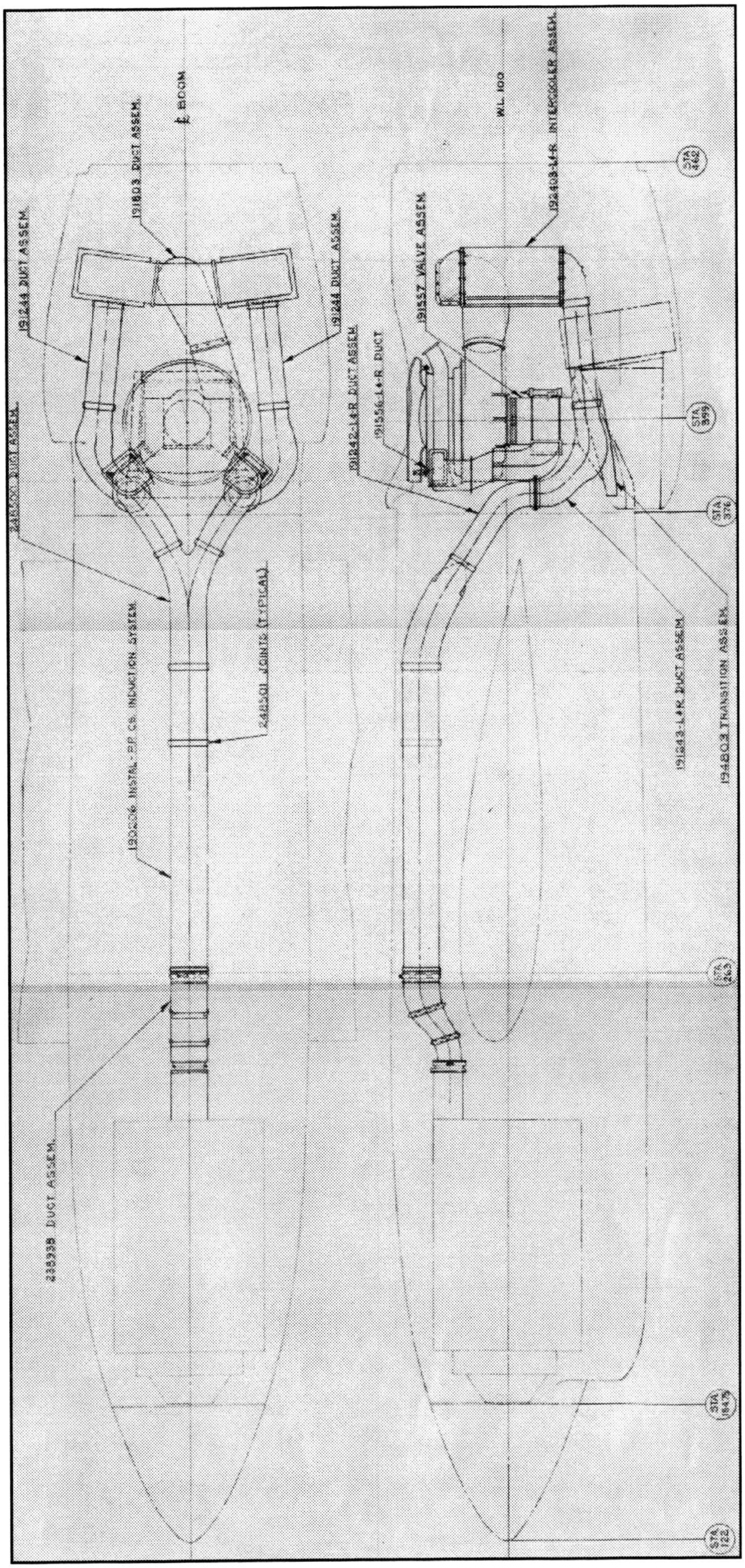

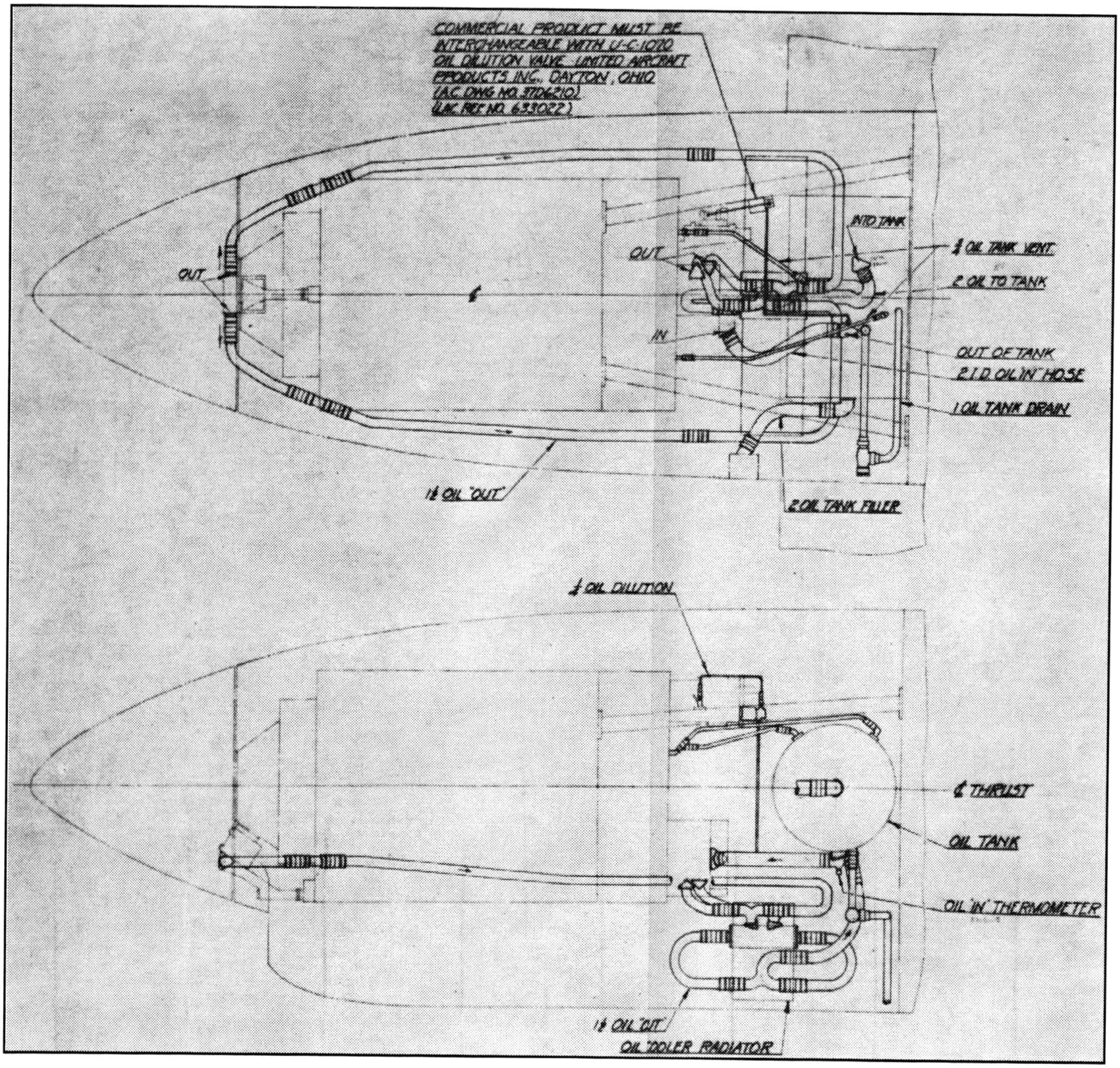
COMMERCIAL PRODUCT MUST BE INTERCHANGEABLE WITH U-C-1070 OIL DILUTION VALVE LIMITED AIRCRAFT PRODUCTS INC., DAYTON, OHIO (A.C. DWG NO. 3706210) (L.A.C REF NO. 633022)
INTO TANK
OIL TANK VENT
OUT
2 OIL TO TANK
OUT
IN
OUT OF TANK
2 I.D. OIL "IN" HOSE
1 OIL TANK DRAIN
OIL "OUT"
2 OIL TANK FILLER
OIL DILUTION
THRUST
OIL TANK
OIL "IN" THERMOMETER
OIL "OIT"
OIL "DOLER RADIATOR"

Glossary

air-cooled: A type of engine in which excess combustion heat is transferred directly to air passing over the engine.
alloy steel: Steel to which small quantities of additional elements such as chromium and molybdenum have been added to the basic iron/carbon mix.
articulated rod: A connecting rod, commonly used in radial engines, that connects all the pistons except the master rod piston in a row to the master rod.
bank: an in-line collection of two or more cylinders, usually cast as one piece.
barrel: An engine construction with multiple cylinders parallel with the power shaft, usually relying on cams, swash plates, or other means to transform linear piston motion into rotary motion of the power shaft.
boost: Additional manifold pressure, usually provided by a supercharger.
boost venturi: A small venturi within the main venturi of a carburetor or fuel injection control unit that produces additional pressure drop for a given flow.
bore: The inside diameter of an engine cylinder.
brake mean effective pressure: The time-averaged pressure acting on an engine piston. Abbreviated BMEP.
BTDC: Before top dead center. Used in references to an engine event, such as ignition, in degrees of crankshaft rotation before the piston reaches the top of its stroke.
built-up crankshaft: A crankshaft construction that typically joins the crankpin to a crankcheek using some mechanical joint such as a clamp or spline.
buried installation: An aircraft engine completely installed within the outline of a wing or fuselage.
coaxial contra-rotating propellers: Two propeller disks, stacked front-to-back along the same axis, and rotating in opposite directions. Used as a means to absorb large amounts with a relatively small propeller, and to eliminate the torque reaction of a single propeller disk.
connecting rod angularity: The angle subtended by the big end of a connecting rod as it follows the crankshaft rotation.
crankpin: The part of the crankshaft to which the connecting rod is attached.
crankshaft counterweights: Mass added to particular locations on a crankshaft to prevent vibration that would be caused by the motion of the connecting rods and pistons.
crankshaft throw: The part of the crankshaft that is offset from the crankshaft main bearings. **detonation:** The fuel-air charge inside the cylinder normally burns in an orderly fashion and at a relatively slow rate. Detonation happens when the fuel-air mixture burns very rapidly, producing extremely high pressure and gas velocity. Pistons and valves quickly overheat and fail. Detonation normally results from low quality fuel, too much supercharging, or early ignition.
direct fuel injection: A type of fuel injection that sprays fuel directly into the cylinder.
displacement: The product of the cylinder area times piston stroke times number of cylinders.
duct: a conduit, usually used to carry air.
Duramold: The Duramold process was invented by Colonel Virginius E. Clark (of Clark "Y" airfoil fame) and consisted of wood and plywood bonded and shaped to its final form using a plastic resin under heat and pressure. Howard Hughes had obtained rights from Fairchild to use the Duramold process on large aircraft, while Fairchild was to continue using it on small aircraft.
dynamic suspension: A scheme to isolate engine vibration from the airframe by mounting the engine on flexible rubber vibration isolators.
dynamometer: An instrument to measure the power produced by an engine.
ethylene glycol: A viscous alcohol that is mixed with water and used as a coolant for liquid-cooled engines.
firing order: The sequence of cylinder firing within an engine.
flat-x: An engine configuration with four banks of cylinders disposed around a central crankcase so that two banks on either side of the crankcase are closer to the horizontal plane than the vertical plane.

forged: A manufacturing process in which a piece of metal that has been heated to the point it is in a plastic state, is hammered into a nearly final shape. The forging process introduces a "grain" to the part which makes it stronger than similar cast parts.
H: An engine configuration with two horizontally-opposed engines are joined to a single crankcase. An H engine typically has two crankshafts that are geared to a single power shaft. The opposed engines may lie in the horizontal plane or the vertical plane.
heat rejection: The ability of a cooling system to dissipate heat from an engine. Only about 25 percent of the heat energy in fuel is converted to power. The remainder must be dissipated via the exhaust or cooling system. In liquid cooled engines, a coolant radiator and an oil radiator are installed to transfer all the waste heat to the air stream.
horizontally-opposed: An engine configuration with a horizontal bank of cylinders on either side of a common crankcase.
horsepower: A measure of power equal to 550 pounds being moved a distance of one foot in one second.
hydraulic lock: A condition in which a cylinder fills with oil or fuel, prohibiting rotation of the engine without damage.
Hyper: A term used to describe a series of high-performance liquid-cooled aircraft engines contracted by the Army during the 1930s and 1940s.
impact tube: A component of a carburetor or fuel control unit that senses the dynamic pressure of airflow.
impeller ratio: The gear ratio between the crankshaft and supercharger impeller.
indicator card: A graph of cylinder pressure as a function of crankshaft rotation angle.
in-line: An engine cylinder configuration with all cylinders extend from the crankshaft in the same plane.
intercooler: A heat exchanger typically placed in the output of a supercharger or turbosupercharger that serves to remove some of the heat of compression.
journal: The part of a crankshaft that is supported by a bearing.
knuckle pin: A steel pin that joins the articulated rod to the master rod.
layshaft: A secondary shaft with gears that mesh with those on a primary shaft. In the Tornado, the primary shaft was the engine crankshaft, with the layshafts being used to transfer crankshaft power to the propeller and engine accessories.
line bored: A manufacturing process that assures all bearings along an axis are accurately positioned and sized.
liquid-cooled: A type of engine in which excess engine heat is transferred to the air through an intermediate liquid that is circulated within the engine and through heat exchangers.
magneto: A self-contained high voltage electrical generator used to provide ignition in aircraft piston engines.
main bearing: A type of bearing that supports a crankshaft at its journals.
manifold: A passage that distributes or collects a fluid.
manifold pressure: The absolute pressure that exists within the intake manifold of an engine.
master rod: A connecting rod that attaches to the crankshaft crankpin. Articulated rods are connected to the master rod.
master rod bearing: The bearing in a master rod that runs on the crankshaft crankpin.
military power: The maximum power an engine is allowed to produce for a short period, typically five minutes, during tactical or combat operation. In the lowest impeller ratio or stage of supercharging, the military power rating is usually the same as the take-off power rating.
nitrided: A process used to form a hard surface on steel parts. The part is subjected to an atmosphere of ammonia gas at a high temperature for a period of hours. Nitrogen from the ammonia enters the steel and forms very hard nitrides that are finely dispersed through the surface.
normal power: The maximum power an engine is allowed to produce continuously.
octane rating: A measure of the anti-detonation qualities of a fuel that compares it to the petroleum isomeric paraffin iso-octane. A fuel with a measure of 100 has the same anti-detonation characteristic as iso-octane. Modern fuels exceed this value through the addition of tetraethyl lead, and are rated by performance numbers, or PN.
one piece master rod: A construction of a master connecting rod without a big end joint. A one piece master rod is used on a built-up crankshaft.

picked out: A plain bearing defect involving loss of bearing material because of chipping at its edges.
piston speed: The time-averaged absolute value of piston travel speed, expressed in feet per minute, for a given engine at a given rpm.
poppet valve: A type of valve that consists of a disk attached to a cylindrical stem that is perpendicular and coaxial with the disk axis. The disk is typically held in contact with a seat by one or more springs. In an internal combustion engine, a cam and connecting suitable mechanism causes the valve to open at the correct point of the engine cycle. The valve is closed by the spring.
pressure carburetor: A device that meters fuel under pressure based on the difference between impact air pressure and boost venturi pressure.
pressure pump: An oil pump that takes oil from the oil tank and forces it through engine lubricating oil passages.
pumping cylinders: Additional cylinders in a two-stroke/cycle engine used to pump the fuel/air mixture into the power cylinders.
pumping losses: Engine energy lost to the necessity of pumping induction air into the cylinders.
pusher: An aircraft configuration that places the engine in front of the propeller.
quill shaft: A shaft that is used as a torsional spring to isolate torsional vibration within a mechanism.
radial: An engine configuration with multiple cylinders arranged around a central crankshaft like the spokes of a wheel.
reduction gear: A gear train that allows the engine to turn at a higher rpm than the propeller.
row: In a radial engine, the collection of cylinders all in one plane around a single crankshaft throw.
scavenge pump: An oil pump that removes oil from an engine and returns it to the oil tank.
second-order counterbalances: Fly weights, usually located co-axial with the crankshaft main bearing journals, that are driven at twice crankshaft speed to counteract the effects of second-order vibration.
second-order vibration: In a radial engine using the master/articulated connecting rod construction, an engine imbalance caused by the inability to completely balance acceleration forces with the additions of counterweights alone.
separate cylinder: An engine construction using multiple cylinders that are individually manufactured and assembled, as opposed to ones cast into a bank.
single-stage, single-speed: A supercharger classification with one impeller (stage) and one crankshaft to impeller drive ratio (speed).
sleeve valve: A type of valve that is formed by a thin cylindrical sleeve that closely surrounds the piston and also closely fits within a larger main cylinder. The sleeve is driven by a crank at its base that imparts both linear and rotary motion to the sleeve. Ports in the sleeve register with intake and exhaust ports near the top of the main cylinder.
specific output: The ratio of engine power to engine displacement.
split bearing: A bearing that is formed by two halves that are split along the longitudinal plane of the bearing.
split master rod: A master connecting rod with a big end that is in two pieces to facilitate assembly upon a one-piece crankshaft. The two halves are held together with two or more bolts.
stellite: An alloy of cobalt, chromium, and tungsten, produced by the Haynes Stellite Company, and used on exhaust valve faces and seats to improve their wear characteristics.
stroke: The distance a piston travels in a cylinder from one extreme to the other.
submerged engine installation: Installation of an engine so that it is completely contained within the outline of the wing or fuselage.
supercharger impeller: A disk with radial vanes that accelerates air that enters near its center and is expelled at its edges.
supercharging: An air pump driven mechanically by the engine that provides additional induction air to the engine to compensate for reduced atmospheric pressure at altitude and/or to produce additional engine power.
takeoff power: The maximum power an engine is allowed to produce at sea-level for a short period of time, typically five minutes.
tetraethyl lead: A compound that raises the resistance of gasoline to detonation.

thermal inertia: A property that allows an engine to be temporarily run at a power setting higher than can be accommodated by its cooling system. The engine parts and coolant temperatures are momentarily raised, but not to a degree that is troublesome so long as the engine does not continue operation at the high power setting for too long a time.
thermocouple: A temperature measuring device that converts heat into electricity.
torquemeter: An instrument that allows direct measurement of engine torque.
torsional vibration: Regular periodic twisting of a structural member such as a crankshaft.
true-motion connecting rod: A type of connecting rod for use in radial engines that eliminates some of the problems with the master/articulated connecting rod scheme by making the motion of each connecting rod symmetrical about the crankshaft center.
turbosupercharger: An air pump connected to a turbine that is driven by engine exhaust.
two-speed propeller reduction gear: Reduction gear capable of two different ratios of crankshaft to propeller speed.
two-speed single-stage: A supercharger classification with one impeller (stage) and two selectable crankshaft to impeller drive ratios (speeds).
two-stage supercharger: A supercharger classification with two impellers (stages).
two-stroke: An engine classification where one power stroke is derived from each two piston strokes
valve tappets: That part of a poppet valve operating mechanism that runs against the cam.
vee: An engine construction with two banks of cylinders arranged about a central crankcase at some angle less than 180°.
vibration dampers: Weights that are loosely attached to a crankshaft in such a way that they oppose crankshaft torsional vibration.
water jacket: A cooling chamber that surrounds the cylinder and cylinder head in a liquid-cooled engine.

Bibliography

Reference Abbreviations and Notes

AFHRA — Air Force Historical Research Agency, Maxwell AFB, Alabama.
AHoF — Aviation Hall of Fame and Museum of New Jersey
NARA — National Archive and Records Administration Branch Depository, Suitland, Maryland.
NASM — National Air and Space Museum (added)
RD — A subgroup of Record Group 342, the "Sarah Clark Collection"
RG — NARA Record Group Number

Abbamont, Gerard A. Letter to E. K. McCord. AHoF Archive File 11-27. February 24, 1990.

Balzer, Gerald H. "Request for Data R40-C." *Journal of the American Aviation Historical Society* 40, no. 4, (Winter 1995): 242–263.

Baugher, Joseph F. *American Military Aircraft.* August 2, 1999. http://home.att.net/~jbaugher1/p49.html.

Baugher, Joseph F. *American Military Aircraft.* September 19, 1999. http://home.att.net/~jbaugher1/p54.html.

Baugher, Joseph F. *American Military Aircraft.* September 6, 1999. http://home.att.net/~jbaugher1/p58.html.

Baugher, Joseph F. *American Military Aircraft.* September 19, 1999. http://home.att.net/~jbaugher1/p69.html.

Baugher, Joseph F. *American Military Aircraft.* March 12, 2000. http://home.att.net/~jbaugher1/p72.html.

Baugher, Joseph F. *American Military Aircraft.* March 12, 2000. http://home.att.net/~jbaugher1/p73.html.

Boyne, Walt. "The Most Elusive Hughes ... The D-2/F-11 Recon Bomber." *Wings* 7, no. 3 (June 1977): 44–55.

NOTE: The following *Case Histories* are divided into three parts, a Summary, a Brief, and a File. Each referring endnote identifies the appropriate *Case History* as well as the part of that case history from which the reference was taken.

Case History of the Hughes D-2, D-5, F-11 Project. Historical Section, Intelligence (T-2), Air Materiel Command. August 1946. AFHRA microfilm reel # A2063.

Case History of the IV-1430 (Continental) Engine. Historical Office, Air Technical Service Command. Wright Field, OH. 1945. NARA RG18, E22, B42. Also AFHRA mircofilm reel # A2073.

Case History of the R-2160 Engine. Historical Division, Intelligence (T-2), Air Technical Service Command. Wright Field, OH. 1945. NARA RG18, E22, B45, File "R-2160." Also AFHRA microfilm reel #A2073.

Case History of the XP-54 Airplane. Historical Office, Air Technical Service Command. November 1944. AFHRA microfilm reel # A2060.

Case History of XP-58 Airplane. Historical Section, Intelligence (T-2), Air Technical Service Command. December 1945. AFHRA microfilm reel # A2064.

"Continental's I-1430 Engine." *Aero Digest* 49 (June 1, 1945): 97.

Correspondence regarding R-2160 restoration. AHoF Archive File 11-27.

Correspondence regarding the R-2160. NARA RG342, RD2631, Box 2266, File "Wright R-2160 Engine (Tornado)."

Correspondence regarding the XP-69. NARA RG342, RD2845, Box 2746.

Daub, Rudolph. U.S. Patent Number 2,355,277. August 8, 1944.

Daub, Rudolph. U.S. Patent Number 2,511,248. June 13, 1950.

Goodman, Willard B. U.S. Patent Number 2,3666,852. January 9, 1945.

Goodman, Willard B. and Taylor, P.B. *Tornado Engine Model Variations, W.A.C. Serial No. 436, Supplement No. 1.* Paterson, NJ: Wright Aeronautical Corporation. October 4, 1939. NASM File BW-730120-01.

Goodman, Willard B. *Summary Report of Flat and Radial Liquid Cooled Engine Design Studies, W.A.C. Serial No. 436.* Paterson, NJ: Wright Aeronautical Corporation. August 30, 1939.
NASM File BW-730120-01.

Hunter, David U. *Supplement No. 1 of Wright Model R-2160 Tornado Engine, 2350 HP at 4000 RPM, Stress Analysis No. 16.* Paterson, NJ: Wright Aeronautical Corporation. January 1940. NASM Wright Field Technical Library Collection, Acc. XXXX-0428, D52.41 Wright/547.

Johnson, Robert E. "Why the Boeing B-29 Bomber, and Why the Wright R-3350 Engine." *Journal of the American Aviation Historical Society* 33, no. 3 (Fall 1988): 174–189.

Kenney, George C. Interoffice Memorandum to Chief, Materiel Division, Washington, DC. July 23, 1941. NARA RG342, RD2442, Box 1985, File "R-4090."

Kneen, Orville H. *Everyman's Book of Flying.* New York: Frederick A. Stokes Company, 1930.

Lundquist, W. G. *Report on Tornado Design Studies, W.A.C. Serial Report No. 458.* Paterson, NJ: Wright Aeronautical Corporation, February 13, 1940. NASM BW-730140-01.

Lundquist, W. G., Hawkins, E.A. and Snider, M. *Summary of 150 Hour Endurance Test on Wright Aeronautical Corporation T-14 Tornado Test Unit Engine, Wright Aero Serial No. 616.* Paterson, NJ: Wright Aeronautical Corporation, November 18, 1941. NASM Wright Field Technical Library Collection, Acc. XXXX-0428, D52.41 Wright/635.

McCord, E. K. Letter to the Aviation Hall of Fame and Museum, February 13, 1990.
AHoF Archive File 11-27.

McFarland, Stephen L. "Higher, Faster and Farther," in *Innovation and the Development of Flight*, ed. Roger D. Launius. College Station: Texas A&M University Press, 1999.

Meyers, B. E. August 5, 1941 Inter-office Memorandum to Assistant Chief, Materiel Division, Wright Field. NARA RG342, RD2442, Box 1985, File "R-4090."

Nahum, A., Foster-Pegg, R.W., and Birch, D. *The Rolls-Royce Crecy.* Derby, England: Rolls-Royce Heritage Trust, 1994.

Nelson, W. J., Czarnecki, K. R., and Lipson, S. *Investigation of the Duct Systems on a ¾-Scale Model of the XP-69 Airplane in the Langley Full-scale Tunnel, Memorandum Report No. L5H18.* Langley, VA: Langley Memorial Aeronautical Laboratory, August 20, 1945.

Pecastaingts, Pierre. "Schneider Trophy," October 23, 2000, http://pgts.decollage.org/cupeng/coupeeng.html, January 19, 2001.

Product Chart, AAF Aircraft Equipment, 1945. AHoF File 12-04.

Restoration Log. AHoF Archive File 12-05.

Schlaifer, Robert. *The Development of Aircraft Engines.* Boston: Harvard University Press, 1950.

Speciale, Mike, NEAM Executive Director, Letter to the author, November 28, 2000.

Stress Analysis of Wright Model R-2160 Tornado Engine, 2350 HP at 4000 RPM, Stress Analysis No. 16. , Paterson, NJ: Wright Aeronautical Corporation, January 1940. NASM Wright Field Technical Library Collection, Acc. XXXX-0428, D52.41 Wright/547.

Swords, A. P. *Report on 50 Hour Development Test per Contract W 535 AC-12632 on Wright Aero Tornado 14, Engine No. 11761, Wright Aero Serial No. 532.* Paterson, NJ: Wright Aeronautical Corporation, December 15, 1940. NASM Wright Field Technical Library Collection, Acc. XXXX-0428, D52.41 Wright/584.

Taylor, P. B. Letter to Assistant Chief, Materiel Division, February 18, 1941. NARA RG342, RD3462, Box 1395, File "R-2170."

Test of 1/20 Scale Model of Republic XP-69 Fighter Aircraft, NACA Report No. 4831. NASM File AR-400490-01.

Vaughn, Guy W., Letter to Paul V. Shields, March 26, 1949. NARA RG342, RD2631, Box 2266, File "Wright Historical Data."

Wagner, William. *Continental! Its Motors and Its People.* Fallbrook, CA: Aero Publishers, 1983.

White, Graham. *Allied Aircraft Piston Engines of World War II.* Warrendale, PA: SAE, 1995.

Whitney, Daniel D. *Vee's For Victory.* Atglen, PA: Schiffer Publishing, 1998.

Wright Tornado Engine Specification No. 617-C. Paterson, NJ: Wright Aeronautical Corporation, September 28, 1939. AHoF Archive File 10-18.

X-1800-A4G Engine Specification No. A-7019. East Hartford, CT: Pratt & Whitney Aircraft, November 21, 1939.

Index

About the Author

Kimble D. McCutcheon was born in Huntsville, Alabama and grew up near there. It was near Huntsville that launch vehicles for the Apollo program and propulsion systems for the Space Shuttle were developed. Steeped in this tradition of propulsion, Kim has been interested in aircraft, engines, and mechanisms for his entire life. He learned to fly at the age of fifteen, paying for his flying lessons by washing and waxing airplanes. Kim financed his higher education by working on airplanes and boats.

While attending the University of Alabama in Huntsville, Kim became enamored with computer hardware and software systems. His professional career was spent developing instrumentation, data acquisition, communications, and control systems for government and industry.

Since his retirement from engineering management in 1993, Kim has divided his time between an aircraft maintenance business, teaching people to fly, and his interest in aircraft engine history. Recently, Kim has been involved in formation of the Aircraft Engine Historical Society. This non-profit organization is dedicated to preserving the written and oral history associated with all types of aircraft engines—from rotaries to rockets. It publishes a quarterly journal and maintains a web presence at http://www.enginehistory.org.

Kim lives in Huntsville with his wife, Shirley.